D0378704

in January

Everything You Wanted to Know
(and a Few Things You Didn't)
About Food in the Grocery Store

Vince Staten

SIMON & SCHUSTER
New York   London   Toronto   Sydney   Tokyo   Singapore

SIMON & SCHUSTER
Simon & Schuster Building
Rockefeller Center
1230 Avenue of the Americas
New York, New York 10020

Designed by Levavi & Levavi, Inc.
Manufactured in the United States of America

1   3   5   7   9   10   8   6   4   2

Library of Congress Cataloging-in-Publication Data

Staten, Vince, date.
Can you trust a tomato in January? : everything you wanted
to know (and a few things you didn't) about food in the
grocery store / Vince Staten
        p.   cm.
Includes index.
1. Groceries—United States.   2. Marketing (Home economics)—United
States.   3. Food Habits—United States.   4. Food industry and trade—
United States.   5. Winn-Dixie Stores.
I. Title.
TX353.S812   1992
641.3'1—dc20
                        93-10046
CIP
ISBN 0-671-76941-3

To Mother and Mammy

Acknowledgments

I'd like to begin by thanking the man or woman who invented junk food; it made this book possible.

Thanks to the public relations staffs at the food manufacturing companies who were kind enough to share their records and files with me; particularly helpful were the folks at RJR Nabisco and Frito-Lay.

I couldn't have written this book without the advice and encouragement of my breakfast club mates: Ronnie Lundy and David Inman.

My sometime collaborator Greg Johnson was—as always—full of suggestions.

Also lending support were Jena Monahan, Sarah Fritchner, Larry Magnes, John Sweetbaby Markel, Tish Wimberly Tate, Jack Kannapell, John Moremen, Margaret Shadbourne, Vicki Sylvester, Sherri Arnett, Debbie Bowditch, and Chris Wohlwend, who would have helped out, but he was sick that day.

Special thanks to my editor, Gary Luke, who invented a new editing term for this book. I expect to see the symbol GFO in future editing textbooks; it means a "get funny opportunity" and it was created for those instances when I took things like non-dairy whitener and partially hydrogenated soybean oil too seriously. Thanks to my agent, Kris Dahl, who came up with the idea for this book and who shepherded it through months of fine tuning, and to her assistant, Gordon Kato, the president of the Spook Jacobs Fan Club.

And last, thanks to my wife, Judy, who is my best resource. When it comes to food, she is a triple threat: Not only can she buy it, she can cook it and she can eat it.

Contents

Can You Trust
a Tomato
in January?

Introduction:

Food and Me

▲▼▲▼▲▼▲▼▲▼▲▼▲▼▲▼▲▼▲▼▲▼▲▼▲▼▲

My grandmother once went twelve years without leaving the family farm. If she needed anything, she sent one of her boys, one of my uncles, to fetch it from the general store three miles up the road. Today my family seldom goes a day without stopping at a grocery of some sort—convenience store, supermarket, hypermarket, or warehouse store.

My grandmother raised eight children on a forty-acre farm that was almost 100 percent self-sufficient. "The only things we ever bought were sugar and coffee," recalls my uncle Luther. The family raised corn and wheat, potatoes and tomatoes. There was a grove of fruit trees. They kept a couple of milk cows for their dairy needs and a couple of slop hogs to eat the garbage and provide bacon and lard. When the harvest was bountiful, they canned and dried for the inevitable bad season. And when the cows had calves, they substituted water for milk in their cooking.

On the other hand, my family is 100 percent dependent on

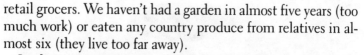

retail grocers. We haven't had a garden in almost five years (too much work) or eaten any country produce from relatives in almost six (they live too far away).

In three generations my family has gone from productive members of the food cycle to gluttonous consumers at the top of the food chain.

My grandmother saw many changes in her lifetime—from horse and buggy to automobile, from country hoedown to cable TV—but nothing changed more than the food supply. She was born in the era of dry goods stores and huckster wagons. By the time she died in 1961, mammoth supermarkets fed by large food conglomerates had taken over the job of supplying food to the masses. These changes paralleled the changes in America; industrialization and urbanization contributed to the evolution of modern food.

I've seen the changes, too. I grew up in the era of neighborhood groceries. "There used to be a grocery store on every block," says my uncle Luther, who ran one for five years back in the late forties. My earliest memory of a food market is Joyner's Grocery, a plain brick store that squatted on a main street between a church and a row of houses. It was the size of a modern convenience store with about the same amount of stock.

Joyner's carried everything: dairy products, canned goods, fresh produce, dry goods; there was even a tiny butcher shop in back. It just didn't carry a lot of anything. There weren't twenty-two varieties of canned corn and a hundred different cereals. There weren't that many different grocery items then anyway, and even if there had been, Joyner's didn't have the shelf space to stock them. Joyner's offered credit and service and home delivery. And Mae and Lewis Joyner knew everybody's name. The place was as much a community center as a grocery store. Neighbors gossiped while plucking items from the shelves and stacking them on the counter, where Mae dutifully totaled them up and added them to the account.

Butch the butcher knew everything that was going on and when he wasn't grinding up hamburger he was spinning tales about the preacher and the spinster. "Preacher was visiting with

Mrs. Tolliver the other day; did you hear about it? She had a dish of nuts on the table and while they talked he kept crunching on 'em. He got ready to leave and apologized because he'd downed every one of them nuts. Told her, 'I'd like to buy you a new can. I don't know what I was doing, they just tasted so good.' 'Oh, never you mind,' Mrs. Tolliver told him. 'My kids bring me cans all the time. 'Cause of my teeth I can't eat 'em, so I just suck the chocolate off.'"

By that time, about 1953, Joyner's was already a dinosaur. We just didn't realize it yet. Supermarkets had already arrived on the scene. In short order they would wipe out all but the most determined of the neighborhood groceries.

The first supermarket in my hometown of Kingsport, Tennessee, was Cut-Rate Supermarket, a store that advertised its main selling point in its name. In 1951 Kingsport had 53 retail grocers; all but 6 were neighborhood stores. Six years later there were 171. The population was exploding in post-war America—the baby boom—and the grocery population was exploding to service it.

But that was pretty much the high water mark. Ten years later, in 1967, the population had doubled, but there were only 173 grocers. Ten years after that the number had headed in the other direction: There were only 139 grocers. Today 67 grocery stores serve a population that has quadrupled since 1951.

There are still a few neighborhood groceries hanging on. But most of the dollar volume has gone to the big chain supermarkets: Food Lion, Food City, and Winn-Dixie. The big stores have efficiency and economy and selection on their side. But for me the grocery store has always been more than a refueling station. It was a place where my mother could ask Mae Joyner about the apples and find out they just came in yesterday, where I could sit on the counter and take in all the smells and sights and sounds and never be out of my mother's sight.

Back then the only people who talked about the health benefits of the food were the old people who were always worried about their regularity. I grew up thinking of food as a pleasure, not a medicine.

In my lifetime supermarkets have come to dominate grocery sales. The supermarket is no longer just a place to buy groceries. It is a deli, a bakery, a video store, and a photo finishing shop. And the food industry that supplies the supermarket is no longer just a marketing machine designed to transport foodstuffs from the place they are harvested to the place they will be consumed. It is a multibillion-dollar megabusiness with laboratories and production centers and government funding. They don't just harvest the food anymore: They fiddle with it, adding a little of this chemical for flavor, a little of that for color, a bit of something else so it can be stored longer.

Convenience is a word you hear frequently when talking to food industry people: It's what the modern cook wants, needs, demands. And gets. But at a price. A price that begins in the farm fields, extends through the packinghouses and processing plants, and concludes in the grocery stores. The modern American diet is nothing if not convenient.

Certainly it is not a new goal. Man has always tinkered with his food, trying to find ways to preserve it, make it taste better, make it cook faster. Europeans suffered through Marco Polo's travel slides because he brought them spices from the Orient to season their bland meals. Travelers to the New World endured the bloating caused by oversalted meat during their long voyages across the Atlantic because they had discovered that storing meat in salt preserved it, meaning they could migrate over great distances without suffering through a month-long all-codfish diet. And bachelors in the 1970s learned to operate toaster ovens without burning themselves so they could cook TV dinners and eat home alone rather than face the humiliation of solo dining in public.

As the years have passed, humans have worked on ways to improve on the natural food supply. They have developed hardier, more disease-resistant strains of vegetables, bred larger fruits, and created varieties of grains with larger yields. They have worked to preserve and store these foods longer. For centuries these food processing progressions could be calibrated in teaspoons and measuring cups. But in the last half of the twentieth century food processing discoveries have exploded, turning a hodgepodge of unrelated techniques into a science, and an industry. And changing forever the American diet.

It's a straight line, if a long one, from the caveman's curdled dinosaur milk to the forever-fresh miracle food called Cheez Whiz. And a necessary line according to Richard Roak, director of the Federal Drug Administration's Center for Food Safety and Applied Nutrition: "Without food additives, we'd live like the caveman."

How much do we really know about our grocery stores and the products we purchase there? We may know where to find the tomatoes in the produce department, but do we know where those tomatoes come from, how they got to the grocery store, and how they differ from the tomatoes we purchased at the grocery only twenty years ago? Many of the products on the grocery shelf were years in development and weeks in production; others were fresh on the vine this time last week.

Only two generations ago all cooks used garden produce, dairy milk, and fresh meat. Now it is a rare cook who takes the time to search out fresh ingredients, not that they can always be found. Meals go straight from the freezer to the microwave. Dinner in ten minutes or less. The changes—from the cream Grandpa skimmed off his cow's milk to lighten his coffee, to Carnation Coffee-mate Non-Dairy Creamer—have been so gradual, so quiet, that nobody has noticed.

And maybe it's time we did.

This book is an investigation into the origins and the development of the foods we take for granted, the foods that are always bulging off our grocery shelves: where they came from and how they got there. It is also about how the grocery stores sell them to us.

This is not a health food book. This is not about the way we should eat. It's about the way we do eat.

I tried to use the best sources available to me at the time of writing. But as anyone who follows the food industry knows, food chemists are always looking for something *new and im-*

proved. My best advice for the reader is to use this book for its intended purpose—a few giggles and a couple of I-didn't-know-thats about the food supply—not as a nutrition guide or as a diet book. If you're looking for more up-to-date information, check the labels.

Men in Paper Hats

▲■▲■▲■▲■▲■▲■▲■▲■▲■▲■▲■▲■▲■▲■▲■

The earliest grocery system was a communistic one—we can admit that now that the Berlin Wall has fallen and the Soviet Union is no longer a union. Cavemen and cavewomen shared what they had. Sometime between Java man and Cro-Magnon man specialization entered the picture. Some cavemen hunted, some gathered, some watched the cave. And one surely set up shop as a grocer.

It didn't take people long to figure out that the food supply was precious and not something to be left to men in paper hats. The first marketplace under government control was in Egypt around 3000 B.C. The Greeks invented the health inspector around 500 B.C. The first markets in this country were Wild West–style trading posts where trappers, Indians, and an occasional greenhorn with gold could swap and shop but not eat. Frontier trading posts didn't offer much in the way of groceries: You were supposed to live off the land back then. These trading posts lived up to their name: trappers trading furs for salt—the first preservative—blankets, and bullets.

As time passed, more and more city folk headed west, settling together in fortified encampments not unlike the modern subdivision, and the trading post gave way to the general store, a retail unit that flourished for two centuries. Here farmers bought what they couldn't raise, usually on credit, and sold what they had in abundance. George Washington purchased everything from provisions to his wooden teeth at the general store. Abe Lincoln worked in a general store in New Salem, Illinois, during his prairie years. The general store seemed like the ideal marketing system: everything under one roof. And it was for the time. But the time was 1850. When roads were nothing more than mud holes and folks couldn't traipse around from the salt store to the blanket boutique.

Then the Industrial Revolution landed on these shores and given a taste of the good life—working in a sweatshop for twelve hours a day seemed like a piece of cake to a farm boy used to working eighteen hours a day in the hot sun—Americans said, what the hey, and made the switch from an agricultural society to an industrial one.

Fewer and fewer people produced food and more and more consumed it. It was a downhill run from living off the fat of the land to worrying about fat in the diet. And as the nineteenth century came to a close general stores began to be pushed out by specialty food markets.

The grocery store as we know it has its roots in a specialty shop. In 1861 George Gilman and George Hartford, two Augusta, Maine, natives opened a tea shop in lower Manhattan at the present-day site of the World Trade Center. Gilman, the more flamboyant of the two business partners, gave it the grandiose name The Great American Tea Company. It was all of 300 square feet. Tea was at the time expensive and the boys' idea was to cut out the long line of middlemen and turn tea into a commodity instead of a luxury.

The first ad for The Great American Tea Company, which appeared in an 1863 issue of the *New York Tribune*, described the tiny outfit as "an organization of capitalists for the purpose of importing tea direct from place of growth and distributing them throughout the United States for one profit only." The idea caught on and by 1865 the company had grown to five stores, all of them in Manhattan. To commemorate the completion of the transcontinental railroad, which was united by the golden spike in 1869, Gilman rechristened the company with an even more grandiose name, The Great Atlantic & Pacific Tea Company. But it quickly acquired a more fluid nickname: the A&P.

In the 1880s Gilman and Hartford added canned milk, butter, coffee, and baking powder to their stock. And slowly the A&P chain began to grow. By 1912 it had 480 stores and $24 million in annual sales. It was the first chain grocery store.

That same year A&P reinvented itself as "The Economy Store," eliminating such services as credit, home delivery, and telephone orders. Price was king. And it worked. A&P benefited from explosive growth over the next decade and a half. By 1927 there were 15,000 A&P stores doing a combined $1 billion a year in sales.

This fixation with low prices was A&P's contribution to the development of the supermarket. But A&P wasn't the first supermarket.

There are a number of rightful claimants to the crown of "first supermarket."

The Piggly Wiggly Store in Memphis, Tennessee
Ward's Grocereteria in Ocean Park, California
Triangle Cash Market in Pomona, California
King Kullen the Price Wrecker in Queens, New York
Big Bear in Elizabeth, New Jersey

These stores were experimenting with the elements that would make the supermarket unique. Dr. Frank Charvat, in *Supermarketing,* places the invention earlier, in 1916, when Clarence Saunders opened his first Piggly Wiggly Store in Memphis. It is generally acknowledged as the first grocery store to incorporate self-service shopping into its design. In its first six months in business, it did $114,000 in gross sales with expenses of only $3,400, enough margin there to make even a Rockefeller smile.

But Ward's Grocereteria in Ocean Park, California, was testing self-service shopping in 1914, a good two years before Saunders. And when the Gerrard brothers opened Triangle Cash Market in Pomona, California, late in 1914, they based it on the Ward's model.

Self-service was not an immediate hit. Customers accustomed to being served by pimply-faced clerks who raced around the store grabbing items off the shelf for them became bewildered when confronted with miles of stocked shelves. The solution for the Gerrard brothers was to arrange all the items in their store in alphabetical order. And in 1917 they changed the name of the Triangle Cash Market to Alpha Beta and tried to sell customers with the slogan "If your child knows the alphabet, he can shop at the Alpha Beta store." There were still problems with this system: Were green beans under *G* or *B*?

Piggly Wiggly and Alpha Beta and Ward's contributed the concept of self-service to the shopping experience, but none of them was the first supermarket. Even the Crystal Palace Market, which opened in San Francisco in 1922 and offered free parking for 4,350 cars, wasn't the first. It didn't offer self-service shopping.

All these early pretenders lacked one element: shoppermania. Shoppers had to make a conscious decision to choose a supermarket in a distant part of town over the local corner grocery.

That was to arrive soon.

The supermarket is a child of the Depression. It was tightened finances that sent shoppers streaming to the supermarkets. Self-service was nice and so was free parking, but nothing was more important than low prices. In 1930 America was ready for the supermarket. Unemployment had skyrocketed that year, from 1.5 million to 7 million. And the one store that intentionally put all the elements of the supermarket together was King Kullen in Queens, New York. King Kullen the Price Wrecker, as it was called, opened March 12, 1930, in an abandoned garage in Queens, New York. It offered self-service, cash and carry, one-stop shopping, and free parking.

Kullen was Michael Cullen, a former A&P employee, whose

concept for a *super* market had been rejected by A&P. Discouraged that his employer of seventeen years had refused to consider his idea, he left to join Montgomery Ward. On his first day at work he met his new boss, a young vice-president. Startled by the youth of his supervisor, Cullen blurted, "By God, do they let kids like you run this business?" The vice-president, unaccustomed to such brashness, wired A&P for Cullen's record and got this reply: "Cullen is an exceptionally good food merchandiser but hard to control."

Cullen and Montgomery Ward soon parted and Cullen set out to raise the capital to open his dream market. His plan was simple: Sell 300 items at cost, 200 items at plus 5 percent, 300 items at plus 15 percent, and 300 items at plus 20 percent.

Cullen and his supermarket profited greatly from his knowledge of the grocery wholesaler side. He would buy drastically reduced merchandise from surplus stocks of canners and other food manufacturers. And he would work every angle to get lower prices from his suppliers. Cullen knew every supplier and every wholesaler in his area; he knew their grocery lines, their stocks, even how they stood with their bankers and when their next bank note was due. "Cash is a magic word and food market operators capitalize that magic," *Business Week* reported. "Chain stores pay most of their bills once or twice a month. Many independent retailers pay when they can. King Kullen pays off all his suppliers every Friday. That happens to be the day before Saturday, and Saturday is pay-day. Why shouldn't a wholesaler grant an extra 5 percent or even 10 percent, a cut of 5 cents, 10

FORKLORE #1

THE BIG PICTURE

Grocery stores in this country chalked up $368.5 billion in sales in 1990. That's larger than the Defense Department's budget, which—you know if you watch *60 Minutes*—is larger than anything on the planet.

cents, 25 cents or even more from his regular per dozen price to make a quick turnover for the dead certainty of getting paid on Friday?" Like anything new, supermarkets were immediately disparaged by the media. *Time* referred to the stores as "cheapies" and *Business Week* labeled them "cheap-jack cash-and-carry depots . . . Cheapness is their motto and the outside, generally including the show windows, is invariably plastered with posters that proudly proclaim the price-wrecking proclivities of the proprietor."

But customers didn't mind.

They didn't mind buying Swift Premium hams for ten cents a pound when traditional grocery stores were charging twelve to fifteen cents a pound. They didn't mind buying Libby's Sliced Pineapple for eighteen cents a can when traditional grocery stores were charging twenty-one to twenty-five cents a can. They didn't mind buying Brookfield Dairy Butter for nineteen cents per pound when traditional grocery stores were charging twenty-one to twenty-five cents per pound.

Within two years King Kullen was doing $150,000 a week in business. And that success was not going unnoticed in the grocery trade.

In December 1932 Roy O. Dawson, who had learned the grocery trade in Memphis under Clarence (Piggly Wiggly) Saunders, astounded New Jersey housewives with the opening of what *Business Week* called "the largest cheapy food market of its kind" on the highway between Newark and Elizabeth, New Jersey.

Dawson called it Big Bear, America's Thrift Center. And his ad in the morning newspaper the day of the grand opening screamed "Big Bear! World's Champion Price Fighter!"

Big Bear was a mammoth grocery bazaar plopped down in an old Durant automobile plant on Freylinghuysen Avenue. Occupying 50,000 square feet of space—more than an acre—it was bigger than most modern supermarkets. There was carnival bunting draped all over the building, glaring neon lights framing the entrance, and hot dog vendors barking their specials from sidewalk carts.

Inside, shoppers were stopped in their tracks by loudspeakers announcing "superspecial Blue Card savings." A sign at the canned fruit juice shouted "Drought Ruins Crops! Prices Rising Up to 50 Percent. Wise Shoppers Buy Now!" There were no display windows, no sales counters; there were just shipping cartons full of merchandise, slashed open and piled high. What Big Bear lacked in ambience it made up for in price.

Big Bear had eliminated all forms of traditional grocery store service: no polite clerks, no credit, no delivery. The only help were the young men who made sure the merchandise was stacked high. Yet shoppers lined up just to get in the parking lot.

In its first three days Big Bear grossed $31,800—a good year for a corner grocery. Old-line grocery retailers sniffed, insisting all the hoopla had attracted curiosity buyers. In its second week Big Bear did $75,500. And the old-line grocers quit sniffing. In its first year Big Bear did $4 million in sales, $166,507 of it profit.

And this was the Depression.

How could Dawson do it? How could he undercut the competition so severely? He had a simple plan: He sold all of his grocery items at cost and made his money by renting out the rest of his space to concessionaires who sold produce, perishables, and household items. Other grocers complained, demanding that manufacturers not sell to the cheapies and hinting that Big Bear might be selling inferior products.

Dawson fought back with a four-page newspaper insert—the first supermarket circular—called *Bear Facts: More Prices Crushed.* The circular snarled: "The bulls can't take it. There are big

---

FORKLORE #2

DIDN'T THERE USED TO BE
A GROCERY STORE HERE? PART I

Kroger's supermarket chain has 2,100 stores and is the nation's largest florist!

---

interests. For convenience's sake, let's call them bulls who are operating with the delusion that they own the public and the public's money."

In the first and second weeks after *Bear Facts* appeared in the paper, Big Bear grossed $80,226 and $91,692. From then on Dawson would spend $10,000 a month on circulars.

The next year, 1933, the first store to actually use the name supermarket opened. It was Albers Supermarket on Seventh Street in Cincinnati and it moved into a building constructed specifically for a supermarket.

Despite the hoopla surrounding King Kullen and Big Bear, the supermarket didn't immediately take off. There were only ninety-six supermarkets in twenty-four cities in 1935.

The large chains resisted this new movement at first. But their customers started telling them something. Even A&P's volume was on the decline.

By 1936 there were 606 supermarkets in 77 cities. And the next year A&P threw in the towel and began converting its 15,000 stores into supermarkets.

But the real supermarket explosion was still to come. The boom years corresponded to the baby boom years, that massive zoom in the U.S. birth rate that began in 1946 as soldiers returned home from the Second World War and that continued for eighteen years.

Collier's magazine reported in 1951 that supermarkets were opening at the rate of better than three a day. And that pace only picked up into the 1960s.

Why the explosion?

Life magazine tried to explain it to its readers in 1958: "In America's suburban, servantless society, the housewife has neither the time nor the opportunity to go from meat market to bakery to confectionery. Instead she herds the children, and sometimes her husband, into the family automobile and drives to the nearest supermarket."

The baby boom ended in 1964, and with it the supermarket

boom. Only 1,500 supermarkets were opened in 1964 and the number has declined ever since.

Although the supermarket per se is not in jeopardy (not in my lifetime), other types of stores that didn't exist in the fifties— convenience stores, warehouse stores, hypermarkets—threaten its dominance.

So far it is more threat than reality. After all, how could a sterile, cheerless warehouse store or a hurry-up-and-buy-your- beer convenience store ever compete with a place that offers pinchable vegetables, squeezable bread loaves, and grown men in paper hats?

The Trip
Begins

▲▼▲▼▲▼▲▼▲▼▲▼▲▼▲▼▲▼▲▼▲▼▲▼▲▼▲▼

It's six-thirty on a Friday night as my wife, Judy, pulls our car into one of the 248 slots in the parking lot of the Winn-Dixie Marketplace on Brownsboro Road in Louisville, Kentucky. This is "our" grocery store, although we certainly don't feel the kinship to it that my mother felt to Joyner's Grocery in the fifties. We used to know the manager, but he was transferred and we don't even know the name of his replacement. We know a few of the clerks by sight and we see many of our neighbors shopping here, but it is "our" store only in the sense that we spend the largest chunk of our food budget here.

There are 364 supermarkets in the Louisville area for us to pick from and we picked this one. It's been our grocery store since we moved here in 1978 and that's a rarity. Our steadfast loyalty isn't uncommon, but about a quarter of all shoppers switch grocery stores each year.

We've stuck with Winn-Dixie for one simple reason: It's the

closest supermarket to our home. It's two miles from our drive-
way to Winn-Dixie's parking lot. There's one other supermarket we frequent with some regularity, a Kroger's superstore, but
it's six miles from our home. And that's the next closest supermarket. Winn-Dixie wins by default.

Convenience is tops on our list because there's a big difference between driving two miles to the grocery and driving six
miles to the grocery. For most people location ranks farther down
the list.

Why do people choose a particular supermarket? A 1991 *Pro-
gressive Grocer* study ranks cleanliness at the top of the list of
things people look for in a grocery store. Nobody wants to buy
food from a dirty store. Second in importance is price tags; shoppers want all items priced. After that, people want a good produce department, "accurate, pleasant checkout clerks," low prices,
freshness dates on products, and a good meat department. You
have to go all the way to tenth place to find the reason we picked
Winn-Dixie: "convenient store location."

Actually, it's unusual that there isn't another supermarket
close to us. You know the old story: A service station builds on
one corner, and soon there are service stations on all four cor-
ners. That's not just a random occurrence. It's an ancient principle of retailing that was first formulated by economist Harold
Hoteling in 1929. He noticed that ice cream vendors on the
beach tended to congregate in the center, rather than stake out
one end or the other. Hoteling called it the principle of minimum differentiation, the tendency of similar businesses to cluster near each other. "It explains why all the dime stores are
usually clustered together, often next door to each other; why
certain towns attract large numbers of firms of one kind; why
an industry such as the garment industry will concentrate in
one quarter of a city. It is a principle that can be carried over
into other differences than spatial differences. The general rule
for any new manufacturer coming into an industry is, 'Make
your product as like the existing products as you can without
destroying the differences.' It explains why all automobiles are
so much alike and tend to get even more alike. . . . It explains

the importance of brand names in commercial, social, and even religious life . . . and . . . it also explains the importance of advertising, for a great part of advertising is little more than an attempt to establish a brand name in the minds of the public."

Hoteling didn't mention it, but another reason was that if a vendor set up on one end of the beach, he would only draw from that end of the beach, but if he set up in the center, he could draw from the entire beach.

This is our major shopping trip of the week, but it is one of four trips our family will make to buy groceries in the next seven days. The other three will be hurry-up stops, to pick up milk and a loaf of bread, to grab a last-minute item for dinner. This is about the national average: 3.4 trips a week to the grocery. We're Americans; we don't know on Monday that we'll be out of orange juice by Friday. How my mother knew that in the fifties baffles me.

We've come on Friday night because it's our only free time together. My wife works during the week. Saturdays are filled with hauling our nine-year-old son to birthday parties and basketball or baseball practice. Not that many years ago, before our son started school and my wife went back to work, Thursday was our grocery day. That was the day of cut-rate specials and coupons in the newspaper.

But things have changed in our family and in many American families. Friday is the new marketing day in America.

But on the other side, Saturday has been creeping up as a major shopping day. It is tied with Thursday as the second most popular marketing day. And as more families become two-income households, it will eventually overtake Friday as marketing day. It already is for people who are employed full-time.

There are many men and women pouring into Winn-Dixie tonight, but the balance is still tilted to the female side. Women shop for food 70 percent of the time. Men do the shopping 17 percent of the time. The rest of the time they do it together.

This trip doesn't really fit in the "both" category. I am accompanying my wife, but I'm not really shopping. Mostly I'm keeping her company. And adding unnecessary impulse items to the cart. There's even been a study about people like me, grocery shopping hangers-on. Most grocery shoppers have someone else with them. Less than half of shoppers shop alone. Sometimes it's a shopping companion for company. About a third of the time it's the kids. But no one takes them for company. It's just easier to drag them along than to find a baby-sitter. And cheaper.

When we enter Winn-Dixie on this Friday evening, my wife has her grocery list scribbled on a notepad. That cherished old habit—the piece of paper magneted to the refrigerator—is still practiced by most shoppers.

But Judy knows she'll add stuff we don't really need. She always does. That makes her part of the crowd: Forty-five percent of all grocery shoppers who make a shopping list deviate from it. And 48 percent of all shoppers make an unplanned impulse purchase of some kind.

Judy also has a wad of coupons in her purse, most of them clipped from the newspaper or the inserts in the Sunday paper. Judy has brought the coupons along with the intent of using them. But if she discovers an unadvertised special on a similar product, she will stuff the coupon back in her purse and go with the lower price.

We aren't religious in our coupon use, not like those coupon-crazy women you see on *PM Magazine*, the ones with basements full of old cereal boxes and milk cartons, the ones who save a hundred dollars a week by redeeming stacks and stacks of coupons. We might save a dollar a week with coupons and that's a little above the national average. A lot of hassle for a little savings.

There are two kinds of coupons: retailer coupons and manufacturer coupons. Retailer coupons promote store brands and nonbrand items such as potatoes and meat. Winn-Dixie prints out retailer coupons along with our grocery tape. We also find retailer coupons in the newspaper. The purpose of manufacturer coupons is to entice shoppers to switch brands. We get them in the Sunday paper and also through the mail. Occasionally we get a coupon in the mail that's addressed to us and not to Occupant. Some of these come from Select and Save, a coupon program run by Computer Marketing Technology of Manhattan. The company uses a computer base to send, say, dog food coupons to dog owners and no one else. Research has shown that these kinds of coupons are redeemed three to five times as often as newspaper coupons.

Today Judy has a Peter Pan coupon and a Hungry-Man coupon that were addressed to us, not Occupant. Peter Pan Peanut Butter is aimed at households with kids, headed by eighteen- to fifty-four-year-olds, in suburban or rural areas. Swanson Hungry-Man Frozen Dinners are aimed at suburban households headed by thirty-five- to fifty-four-year-olds. That's us. But how in the world did they find us? I'll never know.

The first thing I notice when I enter Winn-Dixie is the barricade: checkout stands, bins full of grocery items, everything pushing me in one direction. It's obvious that whoever designed the store wants me to go right. This is a right-handed store. You can go left, but it isn't easy. There is a narrow space between the far-left express checkout and the customer service desk and you can get a cart through there. Our Kroger's store, on the other hand, is a left-handed store.

The grocery store has a very definite idea of how you should navigate your shopping trip.

Studies show that almost everyone who comes in the store will circle around the periphery. That's why high-profit items are placed on the outer walls.

FORKLORE #3

DIDN'T THERE USED TO BE A GROCERY STORE HERE? PART II

You are now more likely to buy a nonprescription drug at a grocery store (40 percent) than at a drugstore (36 percent).

n this Winn-Dixie the items almost everyone buys—milk, s, butter, and bread—are in the back left corner. That means ou follow the traffic flow through the store, you must go past ery aisle to get to these staple items. It's a marketing princi- ie as old as retailing. My father used it thirty years ago in his hardware store. He had rental tools in the rear. I once asked him why he didn't move them to the front, where they would be more convenient for people who came in just to rent a tool. "I don't want them just to rent a tool. I want them to walk past all my merchandise. Maybe they'll see something else they need on the way to the rear and back."

A few years ago bread was in the back right of this store. The theory was that a shopper who ran in just for milk and a loaf of bread would have to circle the entire store, perhaps picking up a few impulse items along the way. But that theory backfired. Regular shoppers knew they would have to make a run through the whole store just to pick up two items, so they stopped in- stead at a convenience store, where the price might be a few cents higher but the walk—and the time for the stop—would be reduced. So in the last remodeling, bread was moved over next to the dairy case.

Still, there are plenty of opportunities in a quick trip to pass tempting impulse items: cookies, videos, potato chips.

This Winn-Dixie's layout is the store plan that has stood for fifty years: produce to the right, meat at the back, dairy items to the left, and grocery items in the middle. But stores also pay attention to image: what the store design is saying to customers. A store that wants to give an image of service may put the cus- tomer service counter first on the shopper route, along with the deli and the bakery. A store that wants to give an image of low prices may put a row of racks with sale items first on the route.

But the choice of the first department does more than set a tone, it also sells. Studies have shown that whatever department is first usually sells about 1 percent more than if it were else- where in the store. Some stores put the bakery up front for the aroma factor. Winn-Dixie has it at the end of the shopping trail. That's because bakery merchandise is fragile and some shoppers may resist carrying a loaf of bread around the entire store for fear it will get squashed accidentally by a can of green beans.

Virtually every supermarket in America owes its basic design to a 1964 U.S. Department of Agriculture experiment. Test supermarkets in three states were redesigned with perishable items—milk, bread, meat, fruit—in the middle aisles and non-perishable items—canned goods, packed foods—on the perimeter. Then 1,300 shoppers were given a fixed amount of time to shop. The results were nothing less than startling: The total of an average shopper's purchases went down 33 percent and the number of items selected fell from eighteen to fourteen. Shoppers were found to be drastically less likely to buy pretzels and coffee, both perimeter items, but more likely to buy citrus fruits, a middle aisle item.

The study concluded that it is the distance a shopper walks, and not the time spent shopping, that determines how much money is spent. Since then stores have been designed to make sure the average shopper travels through the entire store. Staples are distributed around the store so that shoppers have to travel the maximum distance to find them.

Judy's first stop is the cart area to pick up a heavy mesh grocery cart. It's a new model, manufactured by Unarco, Inc., of Oklahoma City, with a small ad for Oreos on the front.

We take the grocery cart for granted. But supermarkets wouldn't be supermarkets if it weren't for these large, convenient mobile market baskets that make extended shopping trips possible—just the sort of experience the supermarket is all about. The Edison of the aisles, the man who invented the shopping cart, was an Oklahoma greengrocer. One day in 1936 Sylvan Goldman, owner of the Standard and the Humpty-Dumpty supermarkets in Oklahoma City, noticed that his customers quit shopping as soon as their wicker baskets got too full or got too heavy for them to carry.

What to do?

Goldman had a stroke of genius, an idea that would ensure the future of the supermarket. He scribbled out the plans for his immodestly titled biography, *The Cart That Changed the World.* That wasn't far from wrong. True supermarketing wasn't possible until shoppers could accumulate mounds of groceries.

Sylvan Goldman invented the grocery cart.

The original X-frame model—two shallow wire baskets mounted on wheels—was three feet tall, two feet long, and a foot-and-a-half wide. And it was olive green to match his store's decor. The baskets were removable so the carts could be folded up and stored.

The carts were assembled by Goldman's maintenance man Fred Young. And at first Young and his crew made only enough for Goldman's stores. That's when Goldman had his second brilliant idea. He would patent the design and sell the carts to other stores. Goldman formed Folding Carrier Company and displayed his invention at the first Super Market Institute Convention in New York in 1937. And did land-office business.

But winning over shoppers was another story. Early supermarket shoppers didn't immediately take to Goldman's newfangled grocery carts. They preferred the market baskets they had carried from home. It took some persuading to get housewives to walk along pushing these goofy-looking carts. Goldman's solution was to pay his employees' wives to walk around the store filling up carts. Other women saw how convenient it was to have a basket on wheels and pretty soon nobody brought their baskets. By the forties the shopping cart was standard equipment everywhere.

There's been little change since then. The baby seat was added in 1947 with the patented Nest-Baskart, which also had an open bottom shelf and a swinging rear panel to allow carts to be nested

---

FORKLORE #4

THIS JUST IN . . .

According to the U.S. Consumer Product Safety Commission, in 1990 there were 33,000 shopping cart accidents. No deaths.

---

and take up less space. About the only modification since then was that little ad on the front of the cart. ActMedia of Darien, Connecticut, pioneered the practice of attaching these junior billboards to shopping carts in 1972. They didn't arrive in our Winn-Dixie until four years ago.

For shoppers the grocery cart is a constant companion from the time they enter the store until they drive out of the parking lot. Some carts have also found their way past the parking lot. A few wind up as laundry carts, go-carts, walkers, even barbecue grills. The average supermarket loses 12 percent of its carts a year to theft. Another 17 percent simply wear out. So, at the average rate of $100 per pop, for a large store that can add up to $5,000 a year to the cost of staying open.

With her purse loaded into the baby seat and her grocery list in her hand, Judy is ready to forge into the grocery store.

It's a daunting challenge: Some 25,855 different items lie ahead, twice as many as there were in the grocery just a decade ago. Wow! How does all this stuff get in the store?

Some of—bread, potato chips, soft drinks, beer, spices—is stocked by the manufacturers. Drive by your local supermarket early in the morning and you'll see bread trucks and potato chip vans out in the parking lot and uniformed men carting cases and cases inside. These are route salesmen who are keeping the shelves stocked, stacked, and neat. At Winn-Dixie these vendors usually stock the shelves on Monday, Wednesday, and Friday.

Other products come in the back door, from the supermarket's warehouse and distribution center. Produce, store brands, and national brands like Nabisco, Kraft, and Procter & Gamble come from this central warehouse.

The least-known link in the food distribution chain is the food broker. Food brokers are independent agents—usually local guys—who bring together the manufacturer and the retailer. A broker may work for as many as 20 different food companies and deal with 500 different food markets in a region.

Kraft General Foods, for example, uses brokers for Kool-Aid, Country Time, and Tang drink mixes because the powdered drink mix market is a seasonal business that requires a lot of activity in supermarkets during a short selling season.

Just because we call this a grocery store doesn't mean everything inside is edible. Whole aisles are now devoted to home and beauty aids, housewares, diapers, greeting cards, flowers. Grocery stores no longer compete just with each other either; they battle virtually every other retail store, from McDonald's to Toys "R" Us, for the consumer's attention and pocketbook.

Our Winn-Dixie remodeled only a couple of years ago, enlarging the size of the store and adding a deli, a cash machine, a frozen yogurt machine, a video section, and a salad bar.

The Kroger's superstore remodeled six years ago and added a book center, a video section, and a sit-down restaurant. Most of the Kroger's renovation was located in the center of the store, pushing grocery items to the edges. I remember the first time I went in after the remodeling. I got lost in greeting cards while looking for a bottle of ketchup. I finally asked a stock boy, "Didn't there used to be a Kroger's here?"

Judy isn't dreading this marketing trip, but she's not exactly looking forward to it either. Neither are many of the other shoppers

---

FORKLORE #5

SO, MR. BISSONETTE, HOW'S BUSINESS?

Curious about how much business your local supermarket does but can't find out because the manager thinks you are nosy—or a spy for the competition—instead of naturally inquisitive?

Then count the number of shopping carts the store has.

According to a 1988 Food Market Institute Report on Key Costs in the Supermarket Industry, the number of carts a store needs depends on its weekly sales volume. A rough rule of thumb is one cart for every $1,000 in weekly business. A hundred carts? Then your store probably does in the neighborhood of $100,000 a week, $5.2 million a year.

streaming in with us. A 1990 University of Michigan study asked people to rank twenty-two daily activities by how much they enjoyed doing them. Food shopping ranked twenty-first, just above housecleaning.

What don't shoppers like about marketing? How could they begrudge an hour spent with men in paper hats? *Adweek's Marketing Week* reported a laundry list of gripes shoppers have about their grocery stores:

The number-one gripe, indicating that American shoppers enjoy griping more than shopping, is "covered up expiration date." A close second—and certainly a legitimate gripe—"old looking meat." Other shopping peeves include "dirty merchandise," long checkout lines, spoiled produce, and "rude employees."

My number-one gripe—those stupid promo announcements over the public address system—finished a distant ninth. I hate it when I hear "Spill in aisle 9" because I have to quit shopping and race over to aisle 9 to see what they spilled.

Produce

▲▲▲▲▲▲▲▲▲▲▲▲▲▲▲▲▲▲▲▲▲▲▲▲▲

The cart is rolling, the right front wheel is wobbling, and that's produce straight ahead as we begin our shopping odyssey.

I don't think I've ever been in a supermarket that didn't steer you immediately into the produce section. That's because produce is the glamour department of the supermarket, where you can actually see and smell and feel the food. Meat is hermetically sealed in plastic; potato chips are locked up in bags. But produce is right there, a sensory experience just waiting. The smell of fresh fruits and vegetables can get your mouth watering. And, as any supermarket manager can tell you, the best customer is a hungry customer.

The modern produce section is a cornucopia of garden delights. Our Winn-Dixie has 243 different items, which is about average for a modern supermarket. Just twenty years ago, the average supermarket produce section had about one-fourth that number.

The increase is due to the fact that the produce department is no longer a slave to the seasons: berries in spring, peaches in summer, apples and pears in autumn. Because when it's winter here, it's summer in Chile and New Zealand. The world is now our garden. And Winn-Dixie is the world's retailer. Imagine that, right here in Louisville, Kentucky, folks.

California and Florida, with their long growing seasons, can supply tomatoes and strawberries for seven, eight, nine months. To fill in, produce suppliers now import seasonal fruits and vegetables from other countries: Guatemala, Venezuela, and, most importantly, Mexico. Almost all winter produce comes from Mexico.

But there's a price to pay for year-round produce. Taste.

The apples and tomatoes of today aren't as tasty as they were thirty or forty years ago. They've been bred for looks and for long hauls in trucks and boats. And for mass-market tastes, which means a blander-tasting product.

Year-round produce has been a success at the cash register. In 1989 produce sales in supermarkets totaled $29 billion. That's a lot of bananas.

The first produce item Judy and I encounter is apples. And that's no accident either. The apple is far and away the most popular item in produce, almost twice as popular as oranges, bananas, lettuce, potatoes, or tomatoes, the runners-up.

And when it comes to broccoli, George Bush isn't the Lone Granger. Broccoli sales aren't even a blip on the produce department's scales. Broccoli accounts for less than 1 percent of produce sales, ranking it down there with asparagus, cauliflower, and summer squash.

Supermarkets break sales down into fresh fruits and fresh vegetables. Vegetables account for 57 percent of sales in the produce department, fruits for 42 percent. (Watermelons, which are a vegetable, are counted as a fruit; and tomatoes, which are a fruit, are counted as a vegetable.)

Despite the widespread availability of packaged snacks, fruit— particularly the apple—remains the most popular snack item, according to a 1989 Roper Organization poll. Of adults who had eaten a snack over the previous twenty-four-hour period, 48 percent reported they had eaten fresh fruit; 42 percent had snacked on a cookie or cake; 33 percent had eaten a salty snack, potato chips or the like; 28 percent had eaten candy; 24 percent had snacked on ice cream; 13 percent had eaten canned fruit; and 7 percent had eaten frozen yogurt.

That adds up to 195 percent, which is probably about right if you figure in overweight people. What it means is that respondents had more than one snack during that twenty-four-hour period. Almost two snacks.

And if you're wondering what the hell the Roper Organization—pollsters to the presidents—is doing investigating snack food, check out the year: 1989. Not an election year. Bills to pay.

Why is the apple such a favorite? Maybe it goes back to Eve. Maybe it goes back to the lunch box. An apple fits in any size lunch box. It requires no special handling, no refrigeration, no special tool to open it, no special implement to eat it. It's easily disposed of, with no environmental consequences. And if you don't get around to eating it today, it won't spoil before tomorrow's lunch. Probably it goes back to that old saying "An apple a day keeps the doctor away." You can bet the American Medical Association didn't coin that saying. Most likely it was some apple growers' trade association.

5 LB. BAG RED DELICIOUS APPLES—$3.98

We're grabbing a bag of Red Delicious apples. The Delicious is the best seller among apples—it accounts for one-fifth of U.S. sales—and for good reason. It looks like an apple should look: vivid red, shiny, blemish-free—"Still Life on Teacher's Desk." It has been bred to look that way.

Taste is another matter. That has been bred out of it. The Delicious is firm and crunchy with only a dot of flavor. It doesn't even leave an apple taste in your mouth. In 1962 curmudgeon A. J. Liebling in *Between Meals* complained, "People have made a triumph of the Delicious apple because it doesn't taste like an apple and of the Golden Delicious because it doesn't taste like anything."

The Delicious is also a triumph of supermarketing: It looks like an apple, even after being hauled 2,000 miles from Washington State, and it smells like an apple. It has everything you could want in a supermarket apple, save taste. But it sells and produce managers are not about to argue with that. Elspeth Huxley, in the 1965 book *Brave New Victuals*, writes: "You cannot sell a blemished apple in the supermarket, but you can sell a tasteless one provided it is shiny, smooth, even, uniform and bright."

Teachers agree that the Red Delicious looks as if it belonged on their desks. In a 1990 *USA Today* poll, teachers' favorite apples were:

Red Delicious	39 percent
Golden Delicious	24 percent
Granny Smith	20 percent
McIntosh	10 percent

About a third of the eighteen pounds of apples each of us eats each year come from Washington. The apple isn't native to central Washington, but the area's climate and soil are perfect for the fruit.

Apples are not picked from spreading trees whose boughs reach to the heavens. The big apple trees are gone. Modern orchards are filled with smaller trees placed closer together. It's a matter of economics. It's more difficult to harvest the crop from big trees. And apple growers want a quick return on their orchards, so they crowd trees together. These orchards aren't filled with apple trees but with apple machines.

As the apples are plucked from the trees and loaded into 900-pound wooden bins, checkers come round to sample each batch. But these checkers don't do it by the bite. They peel off a patch of skin, then insert a stainless steel pressure gauge into the apple. The gauge measures how much pressure it takes to plunge a metal rod into the apple. If it registers 18 to 20 pounds of pressure, it means this lot is filled with good crisp apples, perfect for long-distance shipping. If it measures 12 to 14 pounds, it means the apple is okay to eat now but will turn to mush before it reaches its destination. These apples are sent to the juice factory.

FORKLORE #6

RADIATOR CHARLIE'S MORTGAGE LIFTER

Early seed breeders were fond of unusual names even if the supermarket doesn't carry such hybrids.

Howling Mob corn is a variety of sweet corn that, legend has it, was so wildly popular that screaming townsfolk would mob any farmer who showed up at market with it.

In the early part of this century there were also:

Cut and Come Again beets
Jacobs Cattle and Neckarkonigin beans
Moon and Stars watermelon
Newton Seale cauliflower
Perlzwiebel onions
Purple Calabash tomatoes
Tennis Ball lettuce

But the most unusual had to be Radiator Charlie's Mortgage Lifter tomatoes.

Radiator Charlie was M. C. Byles, a Logan, West Virginia, shade-tree mechanic, who had a fondness for large, red, medium-acid tomatoes, a strain that was hard to find in the early 1930s.

It took him seven years of backyard breeding and selection, but finally Byles developed the perfect plant, a large, succulent tomato with a taste to die for. Or at least mortgage your home for. He began selling seedlings to his neighbors, who told their relatives, who told their neighbors, and soon Radiator Charlie had made enough money to pay off the mortgage on his house.

And that's how Radiator Charlie's Mortgage Lifter tomatoes got their name. Or at least that's the way I heard it.

Mushy apples do not a happy grower make. A 900-pound bin of apples will fetch upwards of $200 from the produce shippers but only about $50 from the juice processors. The checkers also reject apples with a "water core," a buildup of sugar under the skin that makes the apple look translucent. These too will turn to mush and are sent to the juicing plant. In fact, only solid-colored Red Delicious apples go to the produce suppliers. The streaky red ones are shipped to Taiwan, where the red-yellow color is favored.

And not all these apples will be shipped out immediately to supermarkets. The Red Delicious apples that shoppers buy in June have been stored since the previous fall in a low temperature, low oxygen atmosphere.

The days of a produce department offering a bounty of apples—the Jonathan and the Cortland and the Gravenstein and the Baldwin and the Empire and the Ida Red—are gone. Most supermarkets offer only three varieties: the Red Delicious is the red apple; the Golden Delicious is the yellow apple; the Granny Smith is the green apple. That too is a matter of economics: No clerk is required to explain all the different varieties. The store can just fill up the bins. There might be only two varieties in the grocery, the Red Delicious and the Golden Delicious, if studies hadn't shown that 20 percent of shoppers want a tart apple. That's the Granny Smith, an apple originally developed in New Zealand. Washington State now grows more Granny Smiths than does New Zealand.

1 WATERMELON HALF—$4.31

Watermelon in winter?

It's unthinkable, isn't it? But here it is today, sliced open in all its glory, between bananas and vegetables.

When I was in grade school two events served as harbingers of summer: the arrival of baseball cards and the arrival of watermelons. Either event would have small tongues wagging.

Word actually spread faster about watermelons than baseball

cards. Some kid would have to go in the market to discover baseball cards were out. But watermelon! They stacked them out
front, as high as a small boy, sometimes higher if the produce
manager was daring and the truck was full.

Watermelon meant summer was around the corner. The juicy
red vegetable (we *thought* it was a fruit) with the ugly black
seeds only tasted good in summer. There was nothing like a cold
slice of watermelon on a balmy summer evening. Our next-door
neighbor even kept an old refrigerator down in his basement
just for watermelons.

When the first watermelon crop arrived at the grocery, old
men would appear magically, out of nowhere, and make the
rounds of the stack, thumping each melon with a thumb until
the sound of a perfectly ripe watermelon reflected back. It was
a sound only a trained ear could discern. My next-door neighbor Walter Shankel was the best I ever saw at ferreting out perfect melons. He had large fingers anyway and he would go up
and down the rows—ka-thump, ka-thunk, ka-thung—each melon
with its own sound, like a fingerprint. They all sounded the same
to me, but then he would stop in his tracks and wrestle a watermelon from the middle of the pile. That night sitting around
his picnic table, I would swear it was the best I'd ever tasted.

My father tried to invent winter watermelon in the fifties. He
kept a late-summer melon down in a cool, dark corner of the
basement, covered with a ratty old towel. He took it out in January, with much fanfare. He no sooner shoved the knife into the
melon's belly than he shook his head. "No good." He could tell
when the melon gave no resistance that his experiment had
failed. There was no hope for winter watermelon.

Now watermelon is available at my local supermarket year-
round. It's not a treat anymore, something to savor in summer.
It's a commodity, raised in Mexico and trucked to Middle America in the dead of winter.

Watermelon is the big lug of the produce department: lov-
able, friendly, but who wants to tote it home? It's heavy . . . and
messy . . . and full of seeds.

Watermelon growers have been working on all those complaints. The watermelons you see in the grocery today aren't nearly as mammoth as the watermelons of my youth. Farmers have cultivated smaller melons that actually fit inside a refrigerator: one of the major complaints consumers have had over the years about watermelons. Produce departments also sell pre-sliced watermelon in smaller packages. That would have been a sacrilege when I was growing up. You didn't cut into a watermelon until you were ready to eat it. And when you ate it, you ate it all, no messy rinds in the refrigerator.

The seed problem still hasn't been completely solved even though a seedless watermelon was developed in 1948. That's because the seedless watermelon still isn't seedless. There are no black seeds, but there are small, white, edible undeveloped seedpods. Another negative factor is cost; the seedless variety costs ten to fifteen cents more per pound.

Watermelons are making a comeback, thanks to the new winter varieties and to prepackaged slices. Watermelon consumption peaked in 1960 at 17.2 pounds per person. It dropped steadily after that, reaching rock bottom two decades later when Americans ate only 10.6 pounds a person. Consumption has climbed back up lately, though. One reason for the drop was the decline in family size. Smaller families thought the old 25-pounder was too much for them.

Watermelons are one of the few items in the produce section still harvested by hand. That's because they don't ripen after harvesting, so they must be picked at the peak of ripeness, when they are most delicate and prone to breaking and bruising.

And Walter Shankel's thumping method may have worked for him, but modern watermelon experts say the most reliable way to find a ripe melon is by sight. A ripe watermelon should have a healthy sheen and its underside should be yellow.

2 MEDIUM TOMATOES—$1.34

The last thing Judy picks up in the produce department is a bright red tomato. She rolls it in her hand, feeling the texture and firmness, checking the skin. It is a perfect tomato, no black spots, no mushy places.

"When I think about tomatoes in my childhood, this is the way I remember them: big and red and firm," she says.

The truth is perhaps a bit different. The tomatoes of our youth were probably small, red green with a thin skin and dark blemishes.

The modern tomato is a miracle of breeding: It is plump,

FORKLORE #7

LET'S SEE, AT SIXTY-NINE CENTS A POUND THAT FIGURES OUT TO . . .

Ivan Bright and his grandson Jason, both of Hope, Arkansas, hold the world record for the largest watermelon, a 262-pound specimen they grew in 1986. And the world's longest zucchini was raised by Nick Balaci of Johnson City, New York, who grew a 69 1/2-inch Romanian zucchini in 1987.

But they pale next to the world's largest squash. In 1988 Leonard Stellpflug of Rush, New York, trucked a 653 1/2-pound squash to the annual World Pumpkin Confederation weigh-in in Collins, New York. It broke the old squash record by almost 50 pounds. One observer was heard to remark, "It filled up the back of his pickup truck right good." How did he grow it? Stellpflug shrugged. "Well-rotted manure and twenty-two pounds of fertilizer per plant."

Other big produce specimens you couldn't get through the express lane:

World's largest cabbage: 123 pounds by William Collingwood of County Durham, England, 1865.

World's largest lemon: 5 pounds 13 ounces by Violet Philips of Queensland, Australia, 1975.

World's largest tomato: 4 pounds 4 ounces, about the size of a small boy's head, by Charles Roberts of Great Britain, 1974.

---

pinkish red with a thick, bruise-resistant skin. And there's another big difference between today's tomatoes and the tomatoes of thirty years ago. Their garden of origin. Despite the fact
that Louisville is in the heartland, surrounded by rich black-dirt
farmland, the tomatoes at the supermarket today are from Mexico. Virtually all winter tomatoes are imported and 98 percent
of those consumed in Januray, February, and March are from
Mexico.

It's understandable that a January tomato is imported from a
warm weather country. But even if it were tomato season, most
supermarkets would reject the local produce because the tomatoes would be too tender for their distribution centers. All the
produce has to be separated, graded, boxed, and sent to different stores. The local stuff wouldn't survive the handling.

The tomato in Judy's hand was grown in the Culiacán Valley
in Mexico, where the majority of Mexican tomatoes for export
are grown. It was picked only seven days earlier, shaken from the
vine by a giant piece of American-made machinery, a tomato-
puller. It was still green when it left the vine, a good two weeks
ahead of its moment of full ripeness. From the vine it rode a
conveyor belt to the back of the machine, where it fell into a
ten-foot by five-foot by one-foot wooden box with several hundred of its brethren. The box was removed when the machine
came to the end of the row and loaded onto a pickup truck,
where it rode almost two miles to a large block building. Once
inside, the tomatoes were "gassed," treated with ethylene gas.
This is the same gas the tomatoes would give off internally if allowed to ripen on the vine. The artificial gassing "tricks" the
tomatoes into turning red. They don't "ripen," but they do turn
a ripe color.

The unnaturally red tomatoes were then hauled in refrigerated trucks to packinghouses in Nogales, Arizona. It's a day and
a half trip. In Nogales the tomatoes were graded by size, separated into cardboard crates with other tomatoes the same size
and shape, and left on the loading dock, awaiting an east coast
hauler.

The crate holding our tomato was picked up by an independent trucker who hauls tomatoes from Nogales to the eastern seaboard every week. If he ignored Interstate Commerce Commission regulations on how many consecutive hours he can drive—and he often does—it is possible he made the trip from Nogales to the Winn-Dixie distribution center in four days. At the distribution center the tomato boxes were sorted and placed on pallets with other produce heading to our local store. It may have been another twelve hours before the tomatoes were shipped to our store. The stock clerks there didn't get them out on the shelves until the all-night shift. By the time Judy picks up our tomato and examines it this evening, it has been off the vine seven days.

If it had been allowed to ripen at its own rate, our tomato would be black and rotten now. But because it was picked green, gassed, transported in a refrigerated truck, and kept in a refrigerated room in the distribution center, it still looks gorgeous. It is a firm, red fruit, fresh from the fields of Mexico, beautiful to look at, plump and smooth (thanks to inventive breeders), and firm to the knife blade.

That's because home cooks and restaurants have told growers over the years that they want a tomato they can cut without

---

FORKLORE #8

WHAT'S MY LABEL?

You read all the labels; you study up on the additives; you think you know your food. Here's a test.

The label would read like this: Starches, sugars, cellulose, pectin, malic acid, citric acid, succinic acid, anisyl propionate, amyl acetate, ascorbic acid, vitamin A, riboflavin, thiamine, phosphates.

Sounds scary. What is it?

It's a melon.

---

having it juice all over the sandwich. So that's what they've got.

It's a far cry from my father's old homegrown tomato. Once you take a tomato off the vine, it loses its flavor. But you can't ship tomatoes ripe from Nogales. All you'd have would be tomato juice.

So if we want tomatoes in January, we have to take what we can get. A red piece of pulp.

Cereal,
Cake Mixes,
Spices,
Flour,
Sugar

As we round the corner from produce and head down this aisle, Judy checks her shopping list. There at the top, in unmistakable child's scrawl, is The Addams Family Cereal, a special request from our nine-year-old son, Will. He saw a commercial for it earlier this week during a cartoon show.

Judy has seen The Addams Family Cereal in this store before, but she can't remember exactly where it's located in the cereal aisle, so she must scrutinize the cereal display. That's unusual. The average shopper spends just twelve seconds between the time he or she arrives and departs from a food display, and most shoppers handle only the brand they purchase.

As Judy scans the shelves, she sees old-fashioned Kellogg's Corn Flakes and Post Toasties, two of the simple cereals of our youth, on the bottom row. They have almost been squeezed off the shelves by the sugared cereals our children eat: Ghostbusters Cereal, Teenage Mutant Ninja Turtle Cereal, Waldo Cereal.

These cereals, each a sugared wheat product, often floating in a sea of marshmallow bits, is a triumph of marketing. The products literally sell themselves. The kids watch the cartoons and then want the cereal based on the cartoon characters. The cereals seldom live up to the hype. We just pitched a six-month-old box of Robin Hood Cereal into the trash. Will saw the movie, then saw the cereal on TV. When he tasted it, he declared it "okay." We knew that meant another $3.50 down the drain.

The cereals are stacked on three shelves. On the top shelf, at right about my eye level, are Müeslix, All-Bran, Fruit & Fibre— the adult cereals. On the middle shelf, at about my son's eye level, are Froot Loops, Kellogg's Frosted Flakes, Cocoa Puffs. And on the bottom shelf, at stooping level, are the staple cereals: Kellogg's Corn Flakes, Post Toasties.

This confirms a 1990 survey by the Center for Science in the Public Interest, a Washington, D.C., consumer group. They found that the high-sugar cereals were shelved at children's eye level, while the more nutritious brands were on higher shelves. The kids' cereals averaged 44 percent sugar; the adult cereals averaged 10 percent.

While Judy searches for The Addams Family Cereal, I count brands. There are 124 different cold cereals (the industry calls them RTEs, ready-to-eat cereals) in this aisle. Let me repeat that number: 124!

The grocery of my childhood, Joyner's, carried six ready-to-eat cereals: Kellogg's Corn Flakes, Post Toasties, Wheaties, Cheerios, All-Bran (my father's cereal), and Nabisco Shredded Wheat. There weren't many more brands at the time. A decade and a half ago there were only sixty-five ready-to-eat cereal brands.

Why the cold cereal surge? Because demand for cold cereals has surged. In 1970 ready-to-eat cereal sales totaled $659 million. By 1979 that had tripled to $1.9 billion. By 1989 it had increased sevenfold.

There are three distinct types of ready-to-eat cereals: rolled cereals (Kellogg's Corn Flakes and other flaked cereals), puffed cereals (Corn Pops, puffed wheat, puffed rice), and extruded cereals (The Addams Family Cereal and all those other character-shaped cereals).

Rolled cereals are produced by steam cooking whole grain or ground grain, drying it, rolling it out into flakes between high-pressure rollers, and then toasting the flakes in an oven.

Puffed cereals are created by heating whole grains (or extruded shapes in the case of puffed extruded cereals) under high pressure. When the pressure is released the grains expand. The puffing process was developed in the thirties and the original equipment looked like Civil War cannons, thus Kellogg's old Sugar Corn Pops TV commercial extolling the cereal as "shot from cannons."

Extruded cereals are made by forcing dough through a die that shapes it to look like a little Waldo or a little Addams Family person. The shapes are then dried and either toasted or puffed.

That's how you get your basic breakfast cereal.

Many are then sugar coated. The sugar frosting is sprayed on, then dried with heat. In a number of popular cereals vitamins are added. In rolled and puffed cereals the vitamins are sprayed on after toasting or puffing. The cereal is then heat dried. With extruded cereals vitamins and minerals can be sprayed on or added to the dough before cooking. In the case of sugar-frosted cereals vitamins are sprayed on before the sugar coating.

Sometimes the sugar coating spray contains more than just sugar. A cereal such as Froot Loops gets colored in this process. Other cereals may have a flavoring in the frosting.

It seems at least one person in our home eats cereal for breakfast every morning, and we're about average among American households.

Judy has one other cereal on her grocery list besides The Addams Family. But I have a sneaking suspicion we will end up with more than that in our cart.

First in the cart, a box of that old standby, Kellogg's Corn Flakes.

• • •

18 OZ. BOX KELLOGG'S CORN FLAKES—$1.28

A hundred years ago there were no cold cereals. Cold cereal
for breakfast is a twentieth-century phenomenon. When Sir
William Howard Russell, a correspondent in the United States
for the *London Times*, reported back to his countrymen in 1863,
he noted that an American day began with "black tea and toast,
scrambled eggs, fresh spring shad, wild pigeons, pig's feet, two
robins on toast, oysters."

Not a word about cereal.

Breakfast cereal as we know it was popularized around the
turn of the century in Battle Creek, Michigan.

That's right, Battle Creek, a name well known to children of
the 1950s and 1960s. That's where we mailed in our nickels and
dimes and quarters in envelopes stuffed with box tops, and then
we would wait breathlessly at home for the arrival of the Genuine Signal Wrist Flashlight or the Genuine Space Patrol Binoc-
ulars or the Genuine Navy Frogman set ("high pressure propellant
included; ordinary baking powder will work, too").

Battle Creek is synonymous with breakfast cereal and well it
should be, for it was in this small lower Michigan town that the
habit of eating cold cereal for breakfast was born.

There are two names of note here: the Kellogg brothers and
Charles W. Post. They would give us Kellogg's Corn Flakes and
Post Toasties, respectively, the two most popular cold cereals of
the first half of this century. And they knew each other. In fact,
Dr. Post was a patient of Dr. John Harvey Kellogg's, who also
knew Henry Perky, the Denver lawyer who invented shredded
wheat.

How did it happen that cold cereal was invented in a sleepy
little mill town? Well, turn-of-the-century Battle Creek wasn't
a sleepy little mill town. It was a beehive of food faddists, chief
among them Mother Ellen Harmon White, a founder of the
Seventh-Day Adventist Church. The Adventists were an out-
growth of the Millerites, a group led by William Miller, who calculated from the Book of Daniel, chapter 8, verses 13 and 14,
that the second coming of Christ, the Advent, would occur in

1843. When that didn't happen, he recalculated and discovered the date was really October 22, 1844. The movement lost its zip when that day came and went with no big bang. What remained was a small group of the faithful, the Adventists, who reinterpreted the prophecy and settled in to await the Advent. They eventually adopted the name Seventh-Day Adventist at an 1860 conference in Battle Creek.

John Kellogg was raised in the Adventist faith. His family had converted in 1852, the year he was born.

In the 1890s Dr. Kellogg ran the Battle Creek Medical and Surgical Sanitarium, a Seventh-Day Adventist hospital and health spa. Known as the San, it was a popular health resort where the rich and famous—the sick and the hypochrondriac, the fat and the fashionably thin—could plunk down their dollars and be tortured into good health with a diet of bran, grapes, and zwieback. Ronald Deutsch, in *The Nuts Among the Berries*, calls the San "a veritable fountainhead of faddism."

And it was at the San that the venerable cereal Kellogg's Corn Flakes was invented.

The official version of the Kellogg's Corn Flakes story (actually wheat flakes came first) goes like this:

In 1894 Dr. Kellogg and his brother Will Keith Kellogg, the sanitarium's business manager, were experimenting, trying to find a more palatable breakfast food than the hospital's hard zwieback bread, when they chanced upon wheat flakes. They normally made wheat dough and then ran it through granola rollers. But on the day in question the two men were called away in mid-task and when they returned their dough was stale. They decided to run it through the rollers anyway. The result was a small flat flake that when baked was crisp, light, and flavorful.

The official version fails to mention that 1894 was also the year that John Kellogg visited Henry Perky in Denver. And that the two were in competition to develop a flaking process.

Cereal flakes were actually a rearguard action to solve a problem one of the San's patients had with the hospital's regimen. According to Dale Brown in *American Cooking*, "One of Dr. Kellogg's pet theories [was] that people needed to chew dry, brittle food to keep their teeth in shape. . . . Part of the regular regimen of the San had been the endless munching of zwieback, but when a woman complained of breaking a tooth on it, the doctor vowed to produce a substitute."

The result was the wheat flake, which, Brown notes, was "suspiciously like another [food] faddist's product, Granula; Kellogg dared to call his product Granola, but then thought twice about it when he was sued and quickly changed the name to Granose."

Dr. Kellogg introduced the product at the General Conference of Seventh-Day Adventists held in Battle Creek early in 1895.

Granose wheat flakes—and a later variation, the corn flake—were an instant hit with sanitarium patients, who asked to take sacks home. Brown says that in the first year the hospital sold 100,000 pounds of the stuff.

One of those patients was an ulcer sufferer named Charles Post, who spent nine months under Dr. Kellogg's care. This new

FORKLORE #9

THE OTHER KELLOGG

W. K. Kellogg is saluted on every package of Kellogg's cereal with his famous script signature. That was started in the teens when there were so many imitators that Kellogg needed a way to authenticate his brand. Early boxes had the signature followed by "The package of the genuine bears this signature." Or "None genuine without signature." Or "The original has this signature."

W. K.'s brother, Dr. John Kellogg, doesn't have his signature on the box. In fact, he is barely mentioned in official company literature. It's true it was W. K. who was the business genius. But they did develop wheat flakes and corn flakes together. It's just that John Kellogg wasn't interested in building a business empire. He was more interested in, uh, fighting lust. Yes, lust, that base desire that consumed and destroyed many a man and woman in the nineteenth century. Corn flakes were just part of an overall health food diet that would help young people suppress those dangerous desires.

John Kellogg knew it worked; a healthy diet had worked for him. He married in 1879 but never consummated the marriage. Kellogg spent his honeymoon writing *Plain Facts for Old and Young, Embracing the Natural History and Hygiene of Organic Life,* a book that warned against the evils of sex.

He was particularly obsessed with what was politely referred to in those days as self-pollution. The modern medical term is masturbation. In a ninety-seven-page chapter entitled "Secret Vice (Solitary Vice or Self-Abuse)" Kellogg listed thirty-nine (thirty-nine!) "Suspicious Signs" parents should watch for in their children, signs of the secret vice. They included such aberrant behavior as "sudden change in disposition" (number 4), "sleepiness" (number 7), "fickleness" (number 9), "round shoulders" (number 18), "acne" (number 30), and "fingernail biting" (number 31).

It seemed that the only thing a teen could do that wasn't a sign of the secret vice was eat Kellogg's Corn Flakes!

---

breakfast flake did nothing for Post's ulcers, but it sparked his entrepreneurial spirit and in 1895 he developed a coffee substitute, a mushy "breakfast food drink," which he called Postum Cereal Food Coffee. Sounds appetizing, doesn't it?

He formed Postum Cereal Company to manufacture Postum. And in 1897 he added Post Grape-Nuts to his product roster.

It took the Kellogg brothers until 1898 to develop a flaked corn cereal. Wheat was softer than corn and easier to flake. In 1899 Dr. Kellogg formed the Sanitas Nut Food Company, putting thirty-nine-year-old Will Keith Kellogg in charge. These two future breakfast food giants, Kellogg and Post, seemed to move in lockstep after that, into the new century. The problem was, the popularity of their products had brought imitators pouring into Battle Creek. By 1902 there were more than forty cereal factories operating in the shadow of the sanitarium. Everyone, it seemed, wanted to cash in on the well-known Battle Creek name.

In 1904 Post developed his own corn flake cereal, which he called Elijah's Manna. The cereal was well received but not the name. When he tried to export it to Great Britain, the British government refused to register such a religious name for a food, so Post renamed it Post Toasties.

Two years later Will Keith Kellogg went out on his own, founding the Battle Creek Toasted Corn Flake Company on April 1, 1906, with John as his major stockholder. Will's first move was to buy a full-page ad in the July 1906 issue of *Ladies' Home Journal*, paying $4,000 to tout nothing so much as the fact that most of the magazine's readers couldn't buy Sanitas Toasted Corn Flakes.

The ad, in the form of a letter from Will K. Kellogg, president of the Battle Creek Toasted Corn Flake Company, Battle Creek, Michigan, appeared on page 37:

> This announcement violates all the rules of good advertising. Four thousand dollars has been paid for this space to call the attention of the six million readers of the *Ladies' Home Journal* to a new breakfast food, which, at the time this notice is written, less than ten per cent of the readers of the *Journal* can purchase.
>
> We hope that before you read this, we will be in a position to supply all who ask for Toasted Corn Flakes—but today—May 10th—our mills are absolutely unable to meet the demand. We are working night and day—24 hours, 6 days in the week—to fill orders and as fast as brains and skill can build it, new machinery is being added to our equipment.
>
> Another factory has just been purchased, but in spite of this we have orders for nearly a half a million packages which we cannot fill.
>
> We hope by July 1st to be able to fill *all* orders. In the meantime, the great success of Toasted Corn Flakes will no doubt encourage imitators to take advantage of the situation and endeavor to substitute. There is only one genuine Toasted Corn Flakes. If anything else is offered to you—don't judge the merits of Corn Flakes by the substitute.
>
> Toasted Corn Flakes was originated by the food experts

of the Battle Creek Sanitarium. It has a delicious flavor which everyone pronounces as far superior to all breakfast foods on the market. It is this flavor that won the favor.

At the bottom of this page, you'll find an exceptional offer. Get the coupon signed to-day. The dealers you interest will receive prompt attention. And you'll get a cereal supply free for the season.

Kellogg was asking the housewives of America to be his agents, his sales representatives, a unique advertising proposition if there ever was one. Any lady who could convince her grocer to order a case of Toasted Corn Flakes, thirty-six packages, would get a package free. The gamble quickly paid off. And Will Kellogg's little cereal company was soon a major player—the major player— in the breakfast food market.

Kellogg's added All-Bran to its product line in 1916, Rice Krispies in 1928.

The fifties saw the company trying to appeal to grade-school-age baby boomers with a whole slew of sugar-coated cereals: Sugar Corn Pops, introduced in 1950; Sugar Frosted Flakes, in 1952, and Sugar Smacks (later renamed Honey Smacks, then renamed Smacks), in 1953.

Today Kellogg's makes four of the five most popular ready-to-eat cereals: Kellogg's Frosted Flakes, Kellogg's Corn Flakes, Kellogg's Raisin Bran, and Rice Krispies. General Mills' Cheerios complete the top five.

That jibes with our cupboard. Before we left the house I checked our cereal stock. We had one box each of Kellogg's Frosted Flakes, Cheerios, Kellogg's Raisin Bran, Cap'n Crunch, Grape-Nuts, Apple Jacks, Product 19, Kellogg's Cinnamon Mini Buns, and Lucky Charms. Some of what we buy today will be to replenish our stock.

The ninth most popular cereal is Nabisco Shredded Wheat, with only 3 percent of the total market. That astounds me. Of course, it's a big market; 3 percent is still $144 million in sales. Come to think of it, I don't feel so bad for poor old Nabisco Shredded Wheat.

20 OZ. BOX KELLOGG'S FROSTED FLAKES— $3.48

I'd like to be able to figure out how much the sugar coating on Kellogg's Frosted Flakes is costing me. After all, Kellogg's Frosted Flakes is just Kellogg's Corn Flakes with sugar sprayed on it.

Obviously, Kellogg's doesn't want me to make this comparison. The two cereals are sold in different size boxes. To make the comparison, I have to do a little calculating.

Let's see, Kellogg's Frosted Flakes is 17.4 cents an ounce. Kellogg's Corn Flakes is 7 cents an ounce. I am paying 10 cents an ounce for sugar? Maybe we should consider putting this back on the shelf, getting an extra box of Kellogg's Corn Flakes, and eating it with sugar.

Contributing to the high cost of cereals are all those TV commercials. The rule of thumb in the cereal industry is not to spend more than 25 percent of sales on advertising! That's at the wholesale level, but it still means that a goodly chunk of the $3.48 I am paying for this cereal is to support all those Saturday morning commercials.

Kellogg's is not the only company charging a king's ransom for a little spray-on coating. A twelve-ounce box of General Mills' Total has about a penny and a half more in vitamins than its uncoated sister brand General Mills' Wheaties, according to *Nutrition Action Healthletter*, but it costs sixty-five cents more.

20 OZ. BOX GOLDEN CRISP—$2.89

It is unheard of for an established, recognized brand to undergo a name change, but that's exactly what happened in the early eighties when two major national cereal companies had to adjust to a now health-conscious market. Americans were becoming concerned about having too much sugar in their diet and four cereals were unfortunate enough to have the word "sugar" in their names.

Post had Sugar Crisp and Kellogg's had Sugar Frosted Flakes, Sugar Corn Pops, and Sugar Smacks. Both cereal giants solved the problem the same way. But not by reducing the amount of sugar in the cereal. They had actually done that in the early seventies, when a steep sugar price hike sent many food companies looking for an alternative. The alternative was corn syrup.

No, they solved the problem by changing their names. Quietly.

In 1984 Post Sugar Crisp became Golden Crisp, Kellogg's Sugar Frosted Flakes became Frosted Flakes, Sugar Corn Pops became Corn Pops, and Sugar Smacks, having already been renamed Honey Smacks, became Smacks. The only vestige of that simpler time is Sugar Bear, the animated advertising spokesbear for Sugar Crisp, er, Golden Crisp.

24 OZ. BOX GRAPE-NUTS—$2.69

I always thought Grape-Nuts was aptly named: If a grape had nuts, that's what they would look like. For years I thought Grape-

---

FORKLORE #10

SO HOW MUCH FOR JUST THE BOX?

When you buy a box of Kellogg's Frosted Flakes, you aren't paying for just some sugar-coated grain flakes. You are paying for the box; you are paying for transporting the box and contents from Battle Creek to your town; you are paying the salaries of the people who made the dough and sprayed on the sugar; you are even paying for the electricity to run the oven to bake the flakes.

How much? The U.S. Department of Commerce keeps just such statistics.

In 1989 packaging accounted for 10.9 percent of the total food cost. Then labor, transportation, and electricity added another 56.5 percent, with labor accounting for most of that.

Nuts was made from the seeds inside grapes. That was back before botanists developed seedless grapes.

Actually, Grape-Nuts has nothing to do with grapes or nuts. It is made of baked wheat and malted barley. The inventor, Charles Post, named his cereal Grape-Nuts in 1897 because of its nutty flavor and its sweet taste, which he mistakenly believed came from "grape sugar," a common name at the time for dextrose. Grape-Nuts is one of the few cereals in the supermarket with no added sugar.

28 OZ. BOX CHEERIOS—$3.48

Cheerios began life in 1941 as Cheerioats. In 1944 the name was changed to Cheerios.

Cheerios spends more on advertising than any other food product: between $50 million and $99 million annually. No other food item is in that rarefied category. Seventeen food products spend between $25 million and $49.9 million, eight of them in the breakfast food category: Basic 4 cereal, Kellogg's Corn Flakes, Kellogg's Frosted Flakes, Kellogg's Frosted Mini-Wheats, Kellogg's Pop-Tarts, Kellogg's Raisin Bran, Post Grape-Nuts, and Total Raisin Bran.

And what has all that advertising money bought Cheerios? Second place in our hearts.

I helped, in my own small way. I was part of a research group evaluating TV commercials, including one for Cheerios.

It all started with a letter addressed to a "Mr. Staton," which is about as close as anybody ever gets to my name. It was from something called Prevue Studios in Great Neck, New York. The letter began, "Dear TV Viewer, Prevue Studios invites you to attend a screening of special TV programs featuring some new talent, products and commercials. To help evaluate the potential appeal of this new material, the sponsors and people involved in TV production are very interested in obtaining your reactions. So the screening will provide you with a rare opportunity to let them know your opinions about these TV programs and commercials."

It sounded pretty dull—until the last sentence: "In addition, you'll have a chance to win valuable prizes."

I arrived at the Holiday Inn on the appointed evening and was greeted by a bearded fellow who identified himself as the master of ceremonies and handed me a notebook with two questionnaires, a door prize card, and a pen that said "Prevue Studios" on the side. I was ushered into a large meeting room with four giant TV sets. By the time the show started, there were seventy-three of us in the room, about half teenagers and middle-aged women.

Our host, Christopher Bone, appeared on the screen. He was a Ronald Reagan–coiffed veteran of more than 10,000 commercials. In most of them I think he played one of the seven out of ten doctors who recommended the patent medicine. Bone explained what fun we were going to have watching TV shows and commercials and getting a chance to rate them.

But before all that fun began, he wanted us to fill out the basic information on our questionnaires. This would tell Prevue Studios such things as our sex, age, income, and what brand of hemorrhoid cream we used and how often we used it: once a day, three times a week, once a week, once a month, less than once a month (I swear).

Next on the screen was a show called *The Variety Store*. It featured a segment with magician Harry Blackstone Jr., comedian Steve Martin, comic-magician Ballantine, and comedian George Carlin. It was a goofy-looking show. Blackstone appeared to be performing in a Las Vegas lounge, Steve Martin was in some tiny nightclub, Carlin was performing in the round, and God knows where Ballantine was. Between each act there were commercials, just like on a regular TV show.

It was obvious this was not a pilot for a new TV show but what psychologists call a control. From our responses to this show, Prevue Studios would decide if we were a good audience or a bunch of idiots.

Then we were asked to rate the performers. Whom would we rather see as the host of a series, Blackstone or Ballantine? And whom would we rather see starring in his own series, Martin or Carlin? Again this had nothing to do with picking a host for a new series. It was to see if we had average tastes or if we wanted someone stupid like a magician to host a TV show.

Next they had us turn over our questionnaires and write down as many of the seven commercials as we could remember. I could remember four: Kodak, Electrolux, Cheerios, and the hemorrhoid remedy Tronolane, which I called Tonocane. I remembered twice as many as the blue-haired lady in front of me.

Finally we got to the rate-a-new-series part. We watched a half-hour situation comedy called *Dribble*. It was produced by Linda Bloodworth, who created *Designing Women* and *Evening Shade*. It was terrible. I didn't laugh once except at the hemorrhoid commercials in the middle. We rated it and I circled the number that stood for Never Want to See It Again. So did the blue-haired lady in front of me.

I've since learned that *Dribble* was already a failed pilot by the time we saw it. Yet another control. If we liked it, they knew they could throw out all of our data.

Next they wanted us to watch commercials and give our impressions. One commercial turned into a couple of commercials and a couple of commercials turned into four commercials and nine-thirty turned into ten o'clock. We were all tired of trying to remember if the commercial had told us that the hemorrhoid medicine was available in cream, liquid, ointment, or suppository.

Prizes were the last thing on the agenda. They knew we wouldn't leave without a shot at a prize. Before the drawing we were asked to fill out yet another questionnaire to help determine what would be in the fifteen-dollar bag of merchandise that four of us would win.

They gave away four bags of merchandise and teenagers won all of them. And the hemorrhoid cream was an option.

But the rest of us got our reward. For our two hours, we got to keep the Prevue Studios ballpoint pen.

And leave with the knowledge that we had helped shape the future of Cheerios and Tronolane advertising.

20 OZ. BOX THE ADDAMS FAMILY CEREAL— $3.29

Judy finally locates this cereal. It's on the middle shelf, at the end of a long line of sugar cereals. And it's obviously popular, judging by the few boxes left.

Why such a run on a new, unknown, untried cereal?

The chief appeal is the freebie, as witnessed by the fact that half the boxes are missing the Lurch flashlight, a sleek-looking plastic premium attached to the outside of the package.

By my count 65 of the cereals in this aisle have the words "Free inside!" or a similar message emblazoned somewhere on the front of the box. That's 65 out of 124, better than half. Even a presumed health food cereal like All-Bran will often have a free offer.

Premiums have been a sales tool for cereals since 1910, when that first goody—The Jungleland Funny Moving Pictures Book— was packed inside a box of Kellogg's Corn Flakes.

Over the years there have been an assortment of free prizes, from metal license plates of the states to baseball cards. But the most popular have always been miniature toys: plastic gegaws that are simple enough for kids to assemble and sturdy enough to last as long as the box of cereal.

The king of cereal premiums is Manny Winston, a toy designer and manufacturer based in Highland Park, Illinois. He's been creating the little prizes since 1958, when he was working as a graphic artist and a Quaker Oats executive asked him to create a premium toy. Eight hundred million tiny toys later he's still at it.

He often invents five toys a week. He created the Flintstone

Land 'n Sea Vehicle, the Starbot Robot, the Monster Stamp Printer set. His only constraints are his imagination and the five-cent-per-toy maximum most cereal companies are willing to pay. His best creations are the premiums that keep the kids coming back for another, maybe in a different color, or a different design.

If a cereal has a great giveaway inside, it doesn't much matter what the cereal tastes like.

6-COUNT BOX CARNATION CHOCOLATE CHIP BREAKFAST BARS—$2.39

These are called breakfast bars, but my nine-year-old son, Will, takes them in his lunch. Every morning Judy puts one of these in his lunch sack and every night when I unpack his schoolbag, it falls out. In fact, I don't know why we are buying a box of these today. He could just carry the same breakfast bar for the rest of his life. He never eats it.

The operative word in the name here is *bar*. This is really just a candy bar with vitamins sprinkled on it. There are some differences: A candy bar is larger. This is 1.33 ounces. The regulation Mars bars—Milky Way, Snickers, 3 Musketeers—are 2.07 ounces. A candy bar is cheaper—thirty-three cents versus forty cents. And a candy bar tastes better.

You can get almost half of your minimum daily requirement of calcium in one of these. And one-fourth of your niacin and iron. But why would you want to? I think our son would be better off eating a candy bar and drinking a glass of milk. Which is what he does as soon as he gets home from school. While I unpack his schoolbag.

We call this the cereal aisle, but there are a few other necessities on the left side of the row. Like baking soda and pepper. Because it is a right-handed world, supermarkets usually put high-profit impulse items on the right side of an aisle. Which leaves the left side of the aisle for dull, boring items, like baking soda.

32 OZ. BOX ARM & HAMMER BAKING SODA— $1.18

In my lifetime I have seen only two television commercials that left me dumbstruck. One was from the recent AT&T series comparing its long-distance rates with MCI's. In the commercial a man is calling long distance on MCI on one half of the screen and another man is calling long distance on AT&T on the other half. They are supposedly calling the same location and at the bottom of each half of the screen a meter is tallying the cost of each call. When the men hang up, the meters stop to reveal that the MCI call is only a few pennies cheaper than the AT&T call. That's when it struck me: This is an AT&T commercial! AT&T is paying megabucks to the networks to tell me, "Hey, we're not that much more expensive than MCI!" I can't believe that commercial doesn't send people racing to their phones to switch to MCI.

The other unbelievable commercial was a long-running one for Arm & Hammer Baking Soda. The commercial tried to convince me that I should buy Arm & Hammer Baking Soda to freshen my sink drain. All I need to do is pour a box into the drain. That's right: They were saying, "Buy a box of our product and dump it down the drain." What posturing!

5 LB. BAG GOLD MEDAL FLOUR—$2.38

When you think of flour, you think of bread, the staff of life. But since 1972, most of the flour used in home kitchens has been for baking cookies.

5 LB. BAG DOMINO SUGAR—$1.29

We don't need any sugar today, but Domino is on sale for $1.29 for a five-pound bag, with a limit of four per shopper, so we're buying our limit.

Why the limit? Can't Winn-Dixie, with all the marketing muscle of a 1,271-store chain, buy enough to satisfy all of its customers? Or is it such a deal that people will be arriving from all over town, stocking up on sugar, buying enough to last the millennium?

Why the limit? It's a marketing tool, cashing in on the same principles as those at work in the case of the little kid who doesn't want candy until his mother tells him he can't have any. A recent study found that people were more likely to buy an item with a limit than one without. They were also more likely to buy an item when it was marked "limit four" than "limit two."

If America has a sweet tooth—and it must, with the average American consuming 151 pounds of sweeteners a year—then it came by it honestly. Sugar was in part responsible for our victory in the revolutionary war. France entered the war on our side only because it had recently lost the sugar islands of the West Indies to England, crippling France's prosperity. Maybe that's where the expression "Revenge is *sweet*" came from.

There could be an entire chapter in the history of the world called the "Sugar Wars." In *The Fine Art of Food* Reay Tannahill writes, "Sugar became so important to trade that in the 1670s the Dutch yielded New York to England in exchange for the captured sugar lands of Surinam and in 1763 France was prepared to leave England with the whole of Canada, provided she had Guadeloupe returned to her."

I LB. CAN CRISCO—$6.53

Crisco so dominates the shortening market that its name is synonymous with shortening.

Crisco is made from vegetable oils, mainly soybean oil, by the process of hydrogenation, whatever that is. When it first hit the market in 1911, it was a quantum improvement over lard, butter, and other vegetable cooking oils of the time: It didn't pick up odors or flavors from the foods cooked in it, it didn't smoke, and it had a long shelf life.

The name Crisco came from a contest among Procter & Gamble employees. The top two choices were Krispo and Cryst, both of which were meant to imitate the sound shortening makes in a pan. They were combined, sort of, into the name Crisco. Actually, a true combination would be Crispo or Kryst. So P&G hedged the contest a bit.

4 OZ. MCCORMICK GROUND BLACK PEPPER— $2.68

If the three wise men of the East had really been so wise, they wouldn't have wasted their time transporting gold, frankincense, and myrrh to the infant Jesus. Anyone could have purchased those at the local Herod-Mart. They would have loaded up their camels with the most precious commodity in the East, pepper. Yes, pepper, that black spice that produces burning tongues and sneezing fits. There've been times in the past when black pepper was a more stable currency than gold. At various times in history it has been used for dowries, for rent, and even for bribes.

Black pepper comes from the seed of the climbing vine *Piper nigrum*, an East Indian plant now grown in quantity in several countries, chiefly India and Brazil. The seed is harvested before it is ripe and dried in the sun, where it turns black. (White pepper is the interior meat with the black hull removed.) The active ingredient in black pepper is piperine. A mere twenty parts per million is enough to be detected.

Red or cayenne pepper comes from a variety of the plant *Capsicum frutescens*, a relative of the bell pepper, and it is raised in many countries. The active ingredient in red pepper is capsaicin, which is a hundred times more potent than black pepper's piperine.

The burning sensation we call "heat" is not a thermal sensation at all; it is not really heat but an irritation of the nerve endings of the tongue. Over the years food scientists have wrestled with the problem of how to measure that heat. Hot, real hot, and damn hot just weren't descriptive enough. They finally settled on the Scoville units. Pepper is diluted progressively in liquid until it can't be detected by a panel of judges. And that's one Scoville unit.

There are a number of folk remedies used to squelch the burning sensation or "put out the fire." Scientists have been working on that, too. Sensory research has shown that foods that are cold, sweet, salty, or sour help. But don't run and grab a bag of chips, a candy bar, an ice cube, or a slice of lemon. Try a fruity ice cream or frozen yogurt.

2.25 OZ. CAN MCCORMICK LEMON AND PEPPER SEASONING—$1.95

The main ingredient in McCormick Lemon and Pepper Seasoning is—drum roll, please—salt.

AISLE 3

Canned Goods, Coffee, Tea

▞▞▞▞▞▞▞▞▞▞▞▞▞▞▞▞▞▞▞▞▞▞

The big thing now is "No Preservatives!" Every food, it seems, from Top Shelf dinner entrées to Hostess Twinkies, has that phrase plastered on it. Sometimes it's as big as the brand name.

Preservative is as dirty a word as there is in the grocery store— much worse than *sugar* and perhaps on a par with *cholesterol*. But man has been preserving food just about as long as he's been eating it. Hunters couldn't count on a kill a day and gatherers were at the mercy of the elements.

Indians living in the Peruvian Andes 3,000 years ago were the first-known food preservationists, drying potatoes to eat in winter and spring. Colonists heading toward the New World used salt to extend the hatch-life of meat.

Canning was the first major advance in food preservation. The canning process was invented as an adjunct to the conquering process. Napoleon Bonaparte, second in command of France's army in 1795, knew that an army marches on its stomach. As his soldiers went farther and farther afield, conquering new lands, he saw a monotonous diet of salt meat, salt fish, and hardtack take its toll. His soldiers and sailors were dying from scurvy and other diseases associated with dietary deficiencies. He appealed to his country's ruling Directory for better field rations and the Directory responded by offering a prize of 12,000 francs—the equivalent of about $250,000 today—to anyone who could figure out a way of preserving food for transport on military campaigns.

The prize money was so enormous that many men gave up their livelihoods to work at perfecting such a system. Among them was a Paris confectioner named Nicholas Appert. Appert knew the work of Italian priest Lazzaro Spallanzani, who in 1762 had discovered that meat extracts heated in a sealed flask did not spoil, and focused his research in that direction. Fourteen years later he had it: a method of preservation. His solution: Put the food in a glass jar, leaving room for expansion, and stopper the jar with a hand-hewn cork bound to the jar by wire, so as to prevent the cork's blowing out. He then wrapped each jar in a sack and lowered it into boiling water "for more or less time according to [its] nature," as he put it in his 1810 book *The Book for All Households or The Art of Preserving Animal and Vegetable Substances for Many Years*. Using this method, Appert was able to preserve milk, cream, fresh eggs, most fruits and vegetables, even meat and poultry.

In January 1810 Napoleon, now emperor, recognized Appert's contribution and instructed his minister of the interior to award the prize to Appert, "the man who discovered the art of making the seasons stand still."

That same year Englishman Peter Durand made the first tin canister, the "tin can" it would be called, and it was an improvement on the glass jar for preserving foods. The process for making tin cans was tedious; a worker cut the can from a sheet of tinplate, then soldered it by hand. A small hole was left in the top for filling. It would then be covered by a tin disk and soldered again. On a good day a tin worker might make sixty canisters.

Despite the laborious can-making procedure, canning plants

began opening in this country in the 1820s and by the 1840s they were widespread.

The next advance in canning came in 1860 when Isaac Solomon of Baltimore added calcium chloride to the boiling water, raising the boiling point from 212° F. to 240° F. and cutting the processing time from six hours to forty-five minutes. A canning plant that was turning out 2,500 cans a day could now produce 20,000.

Canning innovations peaked with the invention at the turn of the century of the open-top cylindrical can—the "sanitary can." Processors could now include larger pieces of food in the can. Since then the only advances have been in the speed of the equipment.

Canning was a nice invention, but when you've got a garden in the side yard it isn't the necessity it is when you live in a fourth-floor apartment in the middle of a metropolitan area. That's why in 1920 only 3 percent of the fruits and vegetables sold in this country were canned. The vast majority were fresh from the farm. Home gardens were a significant source of the fruits and vegetables in the American diet then.

Today canned goods are almost as popular as fresh produce: Forty percent of the fruits and vegetables sold in this country in 1990 were fresh; 27 percent were canned.

FORKLORE #11

THE SALT SOLUTION

Salt is another one of those nineties naughty words: something we don't want to see listed on our food product's label. But salt is what gives soup its flavor. What to do? Campbell Soup Company, the "M'm! M'm Good!" folks, came up with an innovative solution, one that has been imitated by other food processors. It didn't reduce the salt in the soup. It reduced the serving size listed on the label. It had calculated everything based on a ten-ounce serving. By reducing that to an eight-ounce serving—and changing nothing else—20 percent of the salt in a serving of soup just disappeared.

Each vegetable is canned in a different manner. The intensity of heat used depends on the density of the food, the acidity of the food, the amount of protein and starch in the food, and the number of microorganisms in the food. If you read the ingredients on your can of corn or can of spinach, you'll notice that there are no added preservatives. Canning is considered one of the most benign methods of food preservation. Other time-tested food preservation techniques include salting, pickling, drying, freezing, and smoking. Even the refrigerator, which helps extend the life of food by keeping it cold, is a preservative.

But when people talk about preservatives today, they aren't talking about those methods. They mean chemical preservatives, which have been around at least as long as food processing.

The Food and Drug Administration says a chemical preservative is "any chemical that when added to foods, tends to prevent or retard deterioration, but does not include common salt, sugars, vinegars, spices, or oils extracted from spices, substances added to food by direct exposure thereof to wood smoke, or chemicals applied for their respective insecticidal or herbicidal properties."

The most common chemical preservatives are the benzoates (sodium benzoate, benzoic acid), BHA, BHT, calcium propionate, methylparaben, the nitrites (sodium nitrite, potassium nitrite), propyl gallate, sorbic acid (calcium sorbate, sodium sorbate, potassium sorbate), and sulfur dioxide.

It makes it sound as if you need a degree in chemistry just to eat. In essence, what these preservatives do is prevent mold or dangerous microorganisms from growing in your food. What they do to your body is open to debate. That's why the rush to "No Preservatives" foods.

NO. 2 CAN VAN CAMP'S PORK AND BEANS— $1.29

Originally a No. 2 tin meant there were two pounds of the fruit or vegetable in the tin. But suspicious housewives soon discovered that a two-pound tin of beans might contain only one pound fourteen ounces of beans. In 1910 the designation was changed to No. 1, No. 2, and No. 3 tins and it referred simply to the size of the can and had nothing to do with the weight of the contents.

28 OZ. BOTTLE HEINZ TOMATO KETCHUP— $0.98

Ketchup has changed significantly in my lifetime. The ketchup of my youth was much thinner than today's version. My mother always preferred Hunt's to Heinz because it poured more easily.

What Heinz did to win over consumers in the fifties was nothing short of brilliant. It took a thick, slow-pouring condiment and turned that negative—the incredibly long time it took to get a little ketchup out of the bottle and onto your plate—into an asset with a simple TV commercial: "Heinz. Slowest ketchup in the West . . . East . . . North . . . and South."

It was a play on "the fastest gun in the West," an appellation used liberally in TV westerns of the fifties. And suddenly people began to view ketchup's dense viscosity as a desirable.

Since then Heinz has dominated the ketchup market. About $600 million is spent a year on ketchup. That's about seven 14-ounce bottles of ketchup per person.

Today all ketchups are slow, so slow that David Letterman once demonstrated a way to pour ketchup in a restaurant without making a spectacle of yourself by pounding on the bottom of the bottle: Grip the bottle around the neck and make a complete swing with your arm.

In 1991 the U.S. Department of Agriculture finally recognized that ketchup has changed by altering its ketchup grading standards. By the old standards—established in 1953—ketchup that flowed 9 centimeters in thirty seconds was rated Grade A. (Ketchup is graded on a Rube Goldberg contraption called a "consistometer.") Most popular ketchup brands ooze at 4 to 6.5 centimeters in thirty seconds, so the USDA changed the Grade

A standard to 3 to 7 centimeters in thirty seconds.

What does that mean to you and me? Not much unless we happen to be soldiers, which I don't happen to be. Now Heinz and other thick ketchup makers can bid on government contracts that require strict adherence to USDA grades. If the standards had been changed earlier, Desert Storm soldiers might have been spared the powdered ketchup included in their rations.

FORKLORE #12

AND FIND KETCHUP

Henry J. Heinz, the founder of the world's most successful ketchup company, had four guiding principles to which he credited his business success:

1. Most people are willing to let someone else take over a share of their kitchen operations.
2. A pure article of superior quality will find a ready market through its intrinsic value—if properly packaged and promoted.
3. To improve the product in glass or can, you must improve it while still in the ground.
4. The world is our market.

They aren't quite as catchy as John D. Rockefeller's business principles—get to work early, stay at work late, find oil—but you can't argue with the results. The H. J. Heinz Company today is a $6.6 billion company.

But before you rush out to do a start-up ketchup company, know this: In 1875 Henry Heinz filed for bankruptcy. But he was able to regroup—while his business affairs were still hung up in bankruptcy, his brother and cousin started a new food company with the secret understanding that Henry would become half owner when he emerged from bankruptcy. And the rest is ketchup history.

---

5 OZ. BOTTLE LEA & PERRINS
WORCESTERSHIRE SAUCE—$1.08

Worcestershire sauce wasn't invented so much as it was stumbled upon, in the back of a dank, dark basement. But what a fortunate accident. It seems Lord Marcus Sandys, governor of Bengal, had retired to Ombersley in his native England in 1835. By all accounts it was a pleasant retirement save for one thing. He missed his favorite Indian sauce. He took the recipe to a drugstore on Broad Street in nearby Worcester. The shopkeepers, John Lea and William Perrins, dutifully concocted a large batch, thinking perhaps they might keep some around to sell to other patrons. But one whiff of this vegetable-and-fish mixture convinced them it might best be relegated to the cellar. And that's where it sat, forgotten, for two years. During a cleanup, Misters Lea and Perrins stumbled upon the forgotten sauce,

---
FORKLORE #13

YOU SAY CATSUP, I SAY KETCHUP

Is it *catsup or ketchup?*

In the early years Henry Heinz used both spellings. He didn't settle on *ketchup* until he started advertising the product in the early 1900s. He liked the unique spelling. His prime competitor, J. W. Hunt, pretty much stuck with *catsup* until the 1960s. The last to switch from *catsup* to *ketchup* was Del Monte, which bowed to pressure from its customers and made the change in 1988.

What is it really?

It started out as *ketsiap*, a sauce developed in the seventeenth century by the Chinese. That Oriental concoction would never have made it in this country as a French fry dip: It was a tangy potion of fish entrails, vinegar, and spices. The Chinese used it mainly on fish.

They exported it to Malaya, where it was called *kechap*. It was the Malays who were responsible for the eventual introduction on

continued

these shores. They sold the *kechap* puree to English sailors in the early eighteenth century. Back in England it caught on quickly, but not in its Asian form. English cooks substituted mushrooms for the fish entrails.

The first printed recipe, in Richard Briggs's 1792 cookbook *The New Art of Cookery,* called it catsup. It included tomatoes as an ingredient, a rarity for the time because in Colonial days tomatoes were considered poisonous.

Ketchup didn't enjoy widespread popularity as a food until after 1830. It was on September 26 of that year that Colonel Robert Gibbon Johnson mounted the courthouse steps in Salem, New Jersey, ate a tomato in front of the assembled masses, and didn't die. Undoubtedly someone in the back of the crowd shouted: "I'll bet that would be good on a hamburger!"

---

tasted it, and wow! Lea and Perrins bottled the stuff and began doing a brisk business.

When word reached Lord Sandys about the success of the matured version of his sauce, he returned to demand his 10 percent. To which Mr. Lea is said to have responded: "There's a white coat, m'lord. Put it on and we'll see what can be done." Lord Sandys was not heard from again.

Worcestershire sauce was a local dip until Lea and Perrins's salesmen persuaded the stewards on Britain's passenger ships to add the sauce to their dining room tables. A 1919 ad tried to capitalize on the sauce's popularity on steamships: "Steam takes the place of sail but no sauce has superseded Lea & Perrins, the Original and Genuine Worcestershire. A wonderful liquid tonic that makes your hair grow beautiful." From there it soon became a staple of British dining tables and emigrated to other countries. By the mid-thirties it was an established steak sauce across the Continent. In fact, Lea & Perrins Worcestershire Sauce can be spotted on the table in a September 30, 1938, photo taken in Munich of British prime minister Neville Chamberlain dining with German leader Adolf Hitler, Italian premier Benito Mussolini, and French premier Edouard Daladier. The four talked peace and soon waged war.

The recipe today is pretty much the same as it was in 1835: anchovies layered in brine, tamarinds in molasses, garlic in vinegar, chilies, cloves, shallots, and sugar. The mix is stirred and allowed to sit for, oh, two years. Then the solids are filtered off, preservatives and citric acid are added, and the remaining sauce is bottled. Aside from the preservatives, about the only change in a century and a half has been in the bottle: A drip stopper was added in 1960.

But the major difference between the Lea & Perrins Worcestershire Sauce of a hundred years ago and that of today is it no longer claims to make your hair grow.

13 OZ. FOLGERS COFFEE—$2.08

Lloyd's of London, the British insurance firm famous for insuring Betty Grable's legs and Marilyn Monroe's—uh, what did they insure of Marilyn's?—began as a coffee shop. That's right, a coffee shop. Edward Lloyd attracted a seafaring crowd and as a sideline he began insuring the cargoes and ships of his customers. In 1668 he closed the coffeehouse.

It was just as well. Coffee was declining in popularity, replaced by tea. It was left to us Americans to reinvigorate the coffee market with the invention of the coffee break, a morning respite from routine.

Today coffee is far and away the most popular morning beverage, favored by 47 percent of Americans, according to a survey published in 1990 in *The Wall Street Journal*.

And which coffee do Americans prefer? Folgers.

Folgers regular is more popular than Taster's Choice regular and decaf combined.

Instant coffee was invented by an Englishman named George Washington, who lived in Guatemala around the turn of the century. In 1906 he figured out a way to condense coffee so that all you needed to add to the powder was boiling water.

Instant coffee was a hit and per capita coffee consumption began climbing until it reached a high of 2.45 cups per person per day, in 1962. Then it began a decline that not even the introduction of freeze-dried coffee in 1967 could turn around. By 1977 consumption had fallen to 1.3 cups per person per day. Today the average person drinks 1 cup a day.

I don't drink coffee. Neither do our two sons. That must mean my wife drinks four cups a day.

II OZ. JAR CARNATION COFFEE-MATE NON-DAIRY CREAMER—$1.78

That's an awful-sounding name, isn't it? Non-dairy creamer. It makes it sound as if all you were adding to your coffee was some chemical bleach. Most people must agree with that assessment. Of the coffee drinkers who lighten their coffee, only 17 percent use "non-dairy powders." Most people use milk or half-and-half.

What would coffee be without cream?

Ask the millions of people who don't use cream. Half of all coffee drinkers drink it black.

The most popular non-dairy powder by far is Coffee-mate, which also was the first product of its kind. Coffee-mate can call itself a non-dairy *creamer* because it contains a small amount of sodium caseinate, which is milk protein. Some non-dairy powders don't even have that and have to call themselves "non-dairy whiteners." Yuck.

Well, if it's not cream and it's not a dairy product, then what is it? Vegetable oil and corn syrup; that is, powdered vegetable oil and powdered corn syrup with artificial flavor and a coloring agent, annatto, added so that it doesn't look and taste like what it really is: fat and sugar.

It's colored, flavored powder. It's just a whitener!

It has more calories than milk and more fat than cream. And because it has milk protein, it doesn't offer any advantage for people with milk allergies.

Why are we buying it?

Why does it even exist?

3.4 OZ. BOX JELL-O CHERRY
FLAVOR GELATIN DESSERT—$0.54

In the category of weird foods, Jell-O is right up there with Chee-tos and Pringles. Can you imagine trying to sell it to the pioneers? A food that wiggles? That you can see through?

But Jell-O is a hit. There are other gelatin dessert products—Royal, Knox—but Jell-O is so dominant that it is in danger of becoming a generic term.

Home cooks were making gelatin from animal parts in the eighteenth century, but it wasn't until 1845 that a powdered base for gelatin was invented. The first powdered gelatin mix was patented in 1845 by Peter Cooper, the same guy who invented the Tom Thumb locomotive. Cooper described it as a "transparent, concentrated substance containing all the ingredients fitting it for table use in a portable form and requiring only addition of hot water to dissolve it so it may be poured into moulds and when cold will be fit for use."

But Cooper was too busy driving his locomotive to do anything with his new invention. That was left to a man named Pearl B. Wait, a carpenter in Le Roy, New York. Wait, who had been manufacturing cough medicine and laxative tea out of his home, was looking for a product to sell in the new packaged-foods market. He experimented with Cooper's clear gelatin, adding ingredients until he hit upon a fruit flavor. Wait's wife, Mary, gave the gelatin dessert its distinctive name: Jell-O. The "Jell" part came from *jelly*, which was what it was: animal jelly; and the "O" part was a common ending at the time for food products.

Wait began selling Jell-O in 1897. That's actually a bit charitable: Wait began trying to sell Jell-O in 1897. The former carpenter had no sales or marketing skills and Jell-O sales went nowhere, the same place Wait's bank account was heading. In 1899 he was more than happy to sell his formula and the Jell-O name to another Le Royan, Orator Frank Woodward. That was his name, Orator, not his occupation. Woodward was a go-getter, just what this wiggly dessert needed. He had made several fortunes by the time he bought Jell-O. His first was selling composition target bullets for target practice. Then he invented a medicated cement egg to kill the lice on hens as they were laying or hatching eggs. From there he entered the patent medicine field, selling Kemp's Balsam for Throat and Lungs, Kemp's Laxative, Lane's Tea, Lane's Cold Tablets, liquid Franconia, Sherman's Headache Remedy, and Raccoon Corn Plasters.

His biggest business success, before buying Jell-O, was with Grain-O, a roasted cereal beverage "for those who can't drink tea or coffee." Jell-O became the second product line for his Genesee Pure Food Company.

Jell-O was not an immediate hit for Woodward either, and in 1900 he offered to sell the entire Jell-O business to his plant superintendent for thirty-five dollars. Fortunately for Woodward, the man thought even less of the product than he did.

So Woodward dug in and began a nationwide advertising campaign with a three-inch ad in *Ladies' Home Journal.* A public long accustomed to heavy desserts—pies and cakes—took a fancy to this new light dessert, and Woodward's $336 ad more than paid off. By 1902, sales had topped $250,000 and in 1906 they reached $1 million.

In 1923 Woodward changed the name of his company from Genesee Pure Food Company to the Jell-O Company and on December 31, 1925, Jell-O merged with Postum Cereal Company of Battle Creek to form what would soon become known as General Foods.

The Jell-O of today isn't much different from the stuff Orator Woodward sold: It is still colored sugar water. Jell-O is about 85 percent sugar, 10 percent gelatin, and 5 percent factory-made colors and flavors. The gelatin itself is a flavorless extract of animal hides and bones. That's only a slight improvement upon the gelatin recipes in nineteenth-century cookbooks, which recommended boiling calves feet in water to extract it.

• • •

28 OZ. JAR JIF PEANUT BUTTER—$3.43

There is a Maginot Line for peanut butter. Peanut butter before Jif. And peanut butter after Jif.

Before Jif all peanut butters were dry and clumpy and they stuck to the roof of your mouth; Peter Pan still does.

But Jif was creamy and sweet and went down oh-so-easily. It was also one-fourth cooking oil, but we're getting ahead.

Before Jif was introduced in 1956, peanut butters were very much the way they had been for seventy years, since that day in 1890 when St. Louis physician Ambrose Straub stirred up a batch as a protein substitute for elderly patients with rotten teeth. Not that peanut butter was exactly a new food. Peanut paste had been eaten by Peruvian Indians and African tribes for hundreds of years.

Dr. Straub patented a machine to make peanut butter in 1903. Patent No. 721,651 was granted on February 14, 1903, for a "mill for grinding peanuts for butter." It was launched at the 1904 St. Louis World's Fair.

So peanut butter was born in St. Louis more than a hundred years ago. That means St. Louis is the birthplace not only of the automobile accident (the first one occurred there in 1906 when there were only a handful of cars in the entire state of Missouri) but also of the most popular sandwich spread in America.

The first peanut butters were made from peanuts and salt with perhaps a dab of sugar. The grocery store variety came in large metal containers or wooden tubs, and clerks scooped it out into smaller jars. The early ones weren't very satisfactory. The butter was stiff and hard to spread; the peanut oil tended to separate and rise. It also had a short shelf life, quickly turning rancid.

About the only change in peanut butter during its first half century was the addition of hydrogenated fats to keep the oil from coming to the top and scaring small children.

Enter Jif. Jif was different. It was, the television commercials promised, "creamy smooth." The commercials claimed that "touch of honey" as the magic ingredient that made Jif so creamy.

The FDA claimed differently; it said it was the cooking oil, a half cup—four ounces, 25 percent!—that Procter & Gamble was adding to a sixteen-ounce jar. You need about 3 percent oil to prevent separation. Well, hey, what did they expect from a company that had made a fortune selling Crisco?

The FDA decided that peanut butter, which had never had a federal standard, needed one. "If manufacturers are to be permitted to substitute . . . cheaper vegetable oils for more expensive peanuts . . . the housewife needs safeguards," said commissioner George P. Larrick. The FDA proposed a standard that called for 95 percent ground peanuts in peanut butter. Of course the Peanut Butter Association, which represented the manufacturers who were losing sales to this oily peanut butter, fought the standard. The new definition was published in 1961. Peanut butter makers fought it, taking it all the way to the Supreme Court, so that the standard didn't go into effect until 1971!

This new standard included a compromise with the food industry: Peanut butter need be only 90 percent peanuts. It also allowed salt, nutritive sweeteners, partially hydrogenated vegetable oil (but no lard or animal fat), artificial sweeteners, chemical preservatives, and added vitamins and colors.

Artificial sweeteners? Chemical preservatives? Added vitamins and colors? What happened to the peanuts?

FORKLORE #14

SOME USES GEORGE WASHINGTON CARVER DIDN'T THINK OF

George Washington Carver once served a group of dignitaries a meal that included mock peanut chicken, peanut soup, and peanut ice cream.

He thought of every possible use—no less than 300—for a peanut. Right? Wrong.

Former U.S. senator Barry Goldwater once took peanut butter camping and shaved with it.

Peanut butter is now the most popular use for the peanut. Half of this country's 1.6 million–ton annual peanut crop is used for peanut butter. All other uses—snack peanuts, peanut oil, peanut candy—account for the other half.

Annual consumption of peanut butter in America is well over three pounds per person! The most popular use is for sandwiches: seventy-three percent; 43 percent of those sandwiches are peanut-butter-and-jelly. According to a 1988 Peanut Advisory Board survey, women prefer chunky peanut butter by a margin of 43 to 39. The survey said men have no preference. I do. I prefer creamy. With a touch of honey.

16 OZ. CAN PLANTERS COCKTAIL PEANUTS— $2.48

There are two names that are forever associated with the peanut: Jimmy Carter, the south Georgia peanut farmer who rose to the White House, and George Washington Carver, the Alabama botanist who saved southern agriculture with the peanut

---

FORKLORE #15

THE DOW JIF INDUSTRIAL AVERAGES

Whenever Procter & Gamble chairman Edward Artzt feels a recession coming on, all he needs to do is check his peanut butter sales chart. Since 1970, sales of Jif Peanut Butter have perfectly correlated with recessions. When gross national product was falling and unemployment was rising, as both did in the recessions of 1973–1974 and 1981–1982, Jif sales zoomed.

Artzt explained it this way to *Fortune* magazine: "During lean times, moms turn to other sources of nutrition for their families. Peanut butter offers a good value." And when the economy turns up, kids' fortunes turn down: Mom packs a yucky old tuna fish sandwich in the lunch box.

---

after the Civil War, using his theories of crop rotation.

Carver, as every schoolboy and -girl knows, devised more than 300 uses for the peanut, from instant peanut coffee to peanut soap. In fact, the one peanut dish Carver did not invent was the most popular, peanut butter. It had already been invented.

Before the Civil War, peanuts were something to feed the hogs. A farmer might plant an acre just for that use. Peanuts didn't become a delicacy, a beloved snack, until the 1880s, when P. T. Barnum sold bags of peanuts at his circus shows. About the same time vendors began selling them at baseball games.

People liked these new peanut snacks, but something was missing. That ingredient, salt, was added by Amedeo Obici, an Italian immigrant who ran a fruit stand in Wilkes-Barre, Pennsylvania. He sold roasted peanuts as a sideline and began salting them some time around 1896.

That was about the time he changed his sign to read: Obici, the Peanut Specialist. In 1906, with a partner, he formed Planters Nut & Chocolate Company. Today it is the Planters Life Savers Company, which is now owned by RJR Nabisco.

At the time Obici founded Planters, annual per capita peanut consumption was 2.4 pounds. Today it stands at 9 pounds (!), according to the National Peanut Council.

12 OZ. BOTTLE GREEN GODDESS SALAD DRESSING—$2.68

If you want to know about the creation of Green Goddess

FORKLORE #16

THEN WHAT'S THE FEAR OF PRESERVATIVES?

Arachibutyrophobia is the fear of peanut butter sticking to the roof of your mouth.

Salad Dressing, go to the theater. In particular, go see the William Archer play *The Green Goddess.* In the early twenties the play was such a popular stage attraction that producers were able to lure British stage actor George Arliss to this country to lead a cross-country tour. During the road company's stay in San Francisco Arliss supervised the kitchen staff at the Palace Hotel in the creation of a salad dressing more attuned to his British tastes. From anchovies, may, tarragon, vinegar, and seasonings, the staff created a tangy flavorful dressing. To honor Arliss's contribution and also to recognize the dressing's avocado color, the dressing was dubbed Green Goddess.

16 OZ. JAR SMUCKER'S APPLE JELLY—$2.18

With a name like Smucker's, you gotta be good, right? And Smucker's must be doing something right: It's number one in fruit preserves, jams, and jellies.

The J. M. Smucker Company was founded in 1897 in Orrville, Ohio, by Jerome Smucker—his real name!—who made and sold apple cider and apple butter. In fact, his apples came from trees planted by John Chapman, a.k.a. Johnny Appleseed.

We've been buying Smucker's products since we read a few years ago that they always put a dab more jelly or jam in the jar than the label specifies.

12 OZ. JAR CLASSIC OVALTINE—$3.18

I didn't even know they still made Ovaltine until I saw it on the shelf today, crowded into a small space at the end of the aisle by boxes and boxes of Nestlé Quik and Swiss Miss. This will be my impulse purchase for today. I have to buy a jar to see if it's as bad as I remember it.

Ovaltine may have been the first disappointment in my childhood—either it or one of those heavily advertised toys that looked so great on TV and turned out to be small and flimsy. Ovaltine looked great on TV: The mom stirred it into milk and the kids guzzled it. But at home it tasted cakey and strong, almost like medicine.

Ovaltine is a dinosaur, a product that has outlived its day. Anyone who grew up in the forties or fifties remembers there were plenty of reasons then to buy Ovaltine. It sponsored *Little Orphan Annie* and *Captain Midnight* on radio and *Captain Midnight* on television. Ovaltine offered a secret decoder ring that every kid had to have. It was celebrated in the movie *A Christmas Story*. But now Ovaltine is gone from television, out of sight and almost off the shelf.

Ovaltine was invented in Bern, Switzerland, in the nineteenth century by Dr. George Wander. He named his concoction Ovomaltine, a name that, had it stuck, would have ensured the powdered milk-flavoring wouldn't have lasted long enough to become a dinosaur brand. The name was shortened to Ovaltine when Wander tried to register his trademark in Britain.

Ovaltine was the rage of the world in the thirties. There was the Ovaltiney Club, a secret society for children, that was promoted on Radio Luxembourg. Members got a badge, a rule book and secret code, even their own comic books, *The Chuckler* and *The Dazzler*. There was even a club song, "We Are the Ovaltineys."

Adweek's Marketing Week blames a part of the decline of Ovaltine on its failure to keep up with the times and offer a low-calorie brand. Today the only time you hear anything about Ovaltine is when you rent a *Captain Midnight* episode on videotape. And there is Richard Webb, Captain Midnight, reading a pitch for "nutritious, good-tasting Ovaltine."

24 OZ. BOTTLE MRS. BUTTERWORTH'S SYRUP PRODUCT—$2.28

My mother used to make our waffle syrup when I was a kid. She'd mix brown sugar, butter, and water in a pan and boil it rapidly, stirring while the sugar caramelized. When it reached the right consistency, it was done.

She doesn't do that anymore. Of course, she doesn't make waffles anymore either. If she wants a waffle, she either heads to the Waffle House or pops an Eggo Frozen Waffle into the toaster. And for syrup she reaches into the pantry for a bottle of this stuff.

When I picked this up to put it in the cart, I covered up the bottom part of the label and asked my wife to guess what it was: Mrs. Butterworth's what? "Mrs. Butterworth's Maple Syrup," she said.

And I got to gloat because she hadn't noticed that they had changed the name. This is now called Mrs. Butterworth's Syrup Product. It doesn't contain enough maple syrup to call itself maple syrup.

It does have plenty of syrup, though. The label says it contains .4 percent butter and butter is the third ingredient listed. The ingredients on the label are: corn syrups, sugar syrups, butter, algin derivative, natural flavor, artificial flavor, salt, sodium benzoate and sorbic acid as preservatives, sodium citrate, citric acid, caramel color. Lots of syrup, yeah, but look at all that other stuff in there. Whew! This is a long way from mother's heated sugar water.

It's not really this color; it's not really this flavor; it's not really this consistency.

Lever Brothers, the manufacturer, has added about nine ingredients to my mother's waffle syrup recipe. Those ingredients are called additives. And additives are nothing new. The first patent for an additive was in 1691 for "preserving by liquors and otherwise." In 1783 a patent was issued for a method "for preserving salmon with spices."

But just as additives aren't new, neither were they as rampant in the food of 1691. Additives usage has increased significantly just since the 1950s. In 1955, 419 million pounds of chemical additives were put in our food. Today that number has climbed to 800 million pounds.

Today the federal government recognizes thirty-two different categories of additives, as listed here in the 1974 *Federal Register:*

1. Anticaking agents and free-flow agents: Substances added to finely powdered or crystalline food products to prevent caking, lumping, or agglomeration.
2. Antimicrobial agents: Substances used to preserve food by preventing growth of microorganisms and subsequent spoilage.
3. Antioxidants: Substances used to preserve food by retarding deterioration, rancidity, or discoloration due to oxidation.
4. Colors and coloring adjuncts: Substances used to impart, preserve, or enhance the color or shading of a food, including color stabilizers, color fixatives, and color-retention agents.
5. Curing and pickling agents: Substances imparting a unique flavor and/or color to a food, usually producing an increase in shelf life and greater stability.
6. Dough strengtheners: Substances used to modify starch and gluten, thereby producing a more stable dough.
7. Drying agents: Substances with moisture-absorbing ability, used to maintain an environment of low moisture.
8. Emulsifiers and emulsifier salts: Substances that keep oil and water together in a mix.
9. Enzymes: Enzymes used to improve food processing and the quality of the finished food.
10. Firming agents: Substances added to precipitate residual pectin, thus strengthening the supporting tissue and preventing its collapse during processing.
11. Flavor enhancers: Substances added to supplement, enhance, or modify the original taste and/or aroma of a food, without imparting a characteristic taste or aroma of its own.
12. Flavoring agents and adjuvants: Substances added to impart or help impart a taste or aroma in food.
13. Flour-treating agents: Substances added to a milled flour, at the mill, to improve its color and/or baking qualities, including bleaching and maturind agents.
14. Formulation aids: Substances used to promote or pro-

duce a desired physical state or texture in food, including carriers, binders, fillers, plasticizers, film-formers, and tableting aids.

15. Fumigants: Volatile substances used for controlling insects or pests.
16. Humectants: Hydroscopic substances incorporated in food to promote retention of moisture, including moisture-retention agents and antidusting agents.
17. Leavening agents: Substances used to produce or stimulate production of carbon dioxide in baked goods to impart a light texture, including yeast, yeast foods, and calcium salts.
18. Lubricants and release agents: Substances added to food contact surfaces to prevent ingredients and finished products from sticking to them.
19. Nonnutritive sweeteners: Substances having less than 2 percent of the caloric value of sucrose per equivalent unit of sweetening capacity.
20. Nutrient supplements: Substances necessary for the body's nutritional and metabolic processes.
21. Nutritive sweeteners: Substances having greater than 2 percent of the caloric value of sucrose per equivalent unit of sweetening capacity.
22. Oxidizing and reducing agents: Substances that chemically oxidize or reduce another food ingredient, thereby producing a more stable product.
23. pH control agents: Substances added to change or maintain active acidity or basicity, including buffers, acids, alkalies, and neutralizing agents.
24. Processing aids: Substances used as manufacturing aids to enhance the appeal or utility of a food or food component, including clarifying agents, clouding agents, catalysts, flocculents, filter aids, and crystallization inhibitors.
25. Propellants, aerating agents, and gases: Gases used to supply force to expel a product or used to reduce the amount of oxygen in contact with the food in packaging.

26. Sequestrants: Substances that combine with polyvalent metal ions to form a soluble metal complex, to improve the quality and stability of products.
27. Solvents and vehicles: Substances used to extract or dissolve another substance.
28. Stabilizers and thickeners: Substances used to produce viscous solutions or dispersions, to impart body, improve consistency, or stabilize emulsions, including suspending and bodying agents, setting agents, jellying agents, and bulking agents.
29. Surface-active agents: Substances used to modify surface properties of liquid food components for a variety of effects, other than emulsifiers, but including solubilizing agents, dispersants, detergents, wetting agents, rehydration enhancers, whipping agents, foaming agents, and defoaming agents.
30. Surface-finishing agents: Substances used to increase palatability, preserve gloss, and inhibit discoloration of foods, including glazes, polishes, waxes, and protective coatings.
31. Synergists: Substances used to act or react with another food ingredient to produce a total effect different from or greater than the sum of the effects produced by the individual ingredients.
32. Texturizers: Substances that affect the appearance or feel of the food.

After reading that list, you might logically ask: What aren't they allowed to add? Poison, in short.

Mrs. Butterworth's Syrup Product contains four different categories of additives—flavoring agents, coloring agents, thickeners (algin derivative), and sequestrants (sodium citrate, citric acid)—plus preservatives, which are a class unto themselves.

Why all these additives? There are as many reasons as there are additives, but the National Academy of Sciences created a committee to study additive functions and came up with seven that it considered appropriate uses for additives:

1. To improve or maintain nutritional value
2. To enhance quality
3. To reduce waste
4. To enhance consumer acceptance
5. To improve keeping quality
6. To make the food more readily available
7. To facilitate food preparation

The committee also formulated a list of six inappropriate uses for food additives:

1. To disguise faulty or inferior processes
2. To conceal damaged, spoiled, or inferior goods
3. To deceive consumers
4. To gain functional property at the expense of nutritional quality
5. To substitute for economical, well-recognized good manufacturing processes and practices
6. To use in amounts in excess of the minimum required to achieve intended effects

The committee also insisted that no additive should constitute a health or environmental hazard.

Nowhere does the committee mention—either way—the primary reason for many additives: economics. It's cheaper to use an artificial flavor than it is to use the real thing.

AISLE 4

Canned Fruits, Canned Vegetables, Canned Juices

At the beginning of this aisle, a pleasant woman in an apron offers Judy a free taste of a new healthy snack, Cinnamon Apple Chips. Judy knows the woman—she's one of our neighbors—so she tries them.

"This is really good," she says to the woman. "I bet my kids would like it." Judy adds a bag to our cart. She will later admit to me that she bought the Apple Chips out of guilt. She's done demos—that's what these free sample carts are called—and she knows how long the day can be when no one stops for a taste.

M. C. Snack, Inc., of San Diego is testing Apple Chips in a dozen Louisville grocery stores, all in predominantly white-collar areas. In the past they might have offered these taste demonstrations in all 364 Louisville area supermarkets. But research has shown this to be an inefficient method for introducing products. It costs the company $75 to $100 per store per day for a taste demonstration. Because this product is aimed at a narrow market segment, the company can test it in stores with a high concentration of target shoppers.

Supermarkets like demos because their customers see them as a service provided by the store. For the company it's a relatively inexpensive way to introduce a new product. Not that any way is inexpensive. It costs a minimum of $2 million to introduce a new product. Hormel reportedly spent $25 million to launch Top Shelf dinner entrées. Food companies lose about $20 billion a year to failed new products.

Companies drop this kind of money knowing that nine out of every ten new products fail. They do it because the up side can far outweigh the down side, especially if the new product is a smash hit. Teddy Grahams, those bite-size graham cookies, were a smash hit when they were introduced in 1987, with $150 million in sales the first year.

In addition to passing out free samples, the woman is also distributing coupons redeemable at the cash register. All we need to do is grab a bag of chips from the end-of-the-aisle display behind her.

3-PACK HI-C ORANGE DRINK BOXES—$1.08

We call these drink boxes. The food industry calls them "aseptic packages." I like the name drink box better. "Aseptic package" sounds like something you'd use to wrap a wound, not something you'd drink out of.

Drink boxes are those squat little packs, about the size of a deck of cards, that fit perfectly in a school lunch box. Since its introduction into American supermarkets in 1983, the drink box has practically replaced the thermos bottle. They offer a number of advantages over the thermos: They're lightweight, they're unbreakable, and they're cheap. And because of the aseptic (sterile) packaging, drink boxes don't require refrigeration, so Junior doesn't have to cut a deal with the school janitor.

Aseptic packaging is one of the hot areas in food marketing. It went from zero in 1980—there were no such things as drink boxes in American supermarkets then—to 3 billion boxes sold in 1990. The development of aseptic packaging is so highly regarded in food industry circles that in 1983 members of the Institute of Food Technologists—which doesn't field a football team, even though it calls itself an institute—voted it the number-one food innovation of the last fifty years. That puts it ahead of the microwave oven, ahead of freeze-drying, ahead even of Pringles stackable Potato Crisps.

What is aseptic packaging? *Aseptic* means "free of pathogenic organisms." *Pathogenic* means "capable of causing disease." So an aseptic package, a drink box, is a sterile container.

The liquid is sterilized at a high temperature, then poured into the sterile container, usually the familiar paper-and-plastic-coated drink box, and sealed. Because the product is sterile, there is no need for refrigeration. Because the juice was quick-heated, it retains much of its flavor and nutrients, more than in traditional canning methods.

Coca-Cola is the biggest user of drink boxes. In 1990, 35 per-

<hr>

FORKLORE #17

HOW MUCH JUICE IS IN YOUR JUICE?

Juice cocktails and juice punches and juice drinks are not the same thing as juice. A better name for these products might be "expensive sugar water." In 1980 the FDA tried to require juice makers to tell the juice content on the labels of juice drinks and cocktails and punches. But the cranberry juice industry—read Ocean Spray—fought the change and it still hasn't been instituted.

Here are the juice contents of five common drinks:

Del Monte Fruit Blend	50 percent
Minute Maid Fruit Punch	10 percent
Ocean Spray Cranberry Juice Cocktail	27 percent
Tropicana Pineapple Grapefruit Beverage	10 percent
Welch's Grape Juice Beverage	50 percent

cent of all the drink boxes produced, about 1 billion, were used for Coca-Cola products such as Hi-C fruit drink. In fact, drink boxes rejuvenated Hi-C sales after Coca-Cola switched it from those clunky forty-six-ounce cans to single-serving drink boxes in 1983. Hi-C has seen double-digit sales increases every year since.

Aseptic packages, both boxes and foil pouches, are made of six tightly compressed layers—usually of plastic, paper, and aluminum. The inside layer is made of polyethylene plastic. It prevents spoiling, contamination, and leaking. Next comes a second polyethylene layer, this one for bonding the inner liner to an aluminum foil layer. This aluminum foil protects against light and oxygen, preventing spoiling and taste change. Next comes another layer of polyethylene plastic. Then a layer of paper to give the box shape and strength. The top layer is a water-resistant polyethylene plastic coating to keep the package dry. And when they are empty you can just throw them away, even if it isn't politically correct to do so.

3-PACK GATORADE DRINK BOXES—$1.08

One of the silliest products of the eighties was something called "sports gum." My kids all had to have it before soccer. It was as if it were some magical potion that assured a win. It didn't, of course. And that may have been the problem.

Sports gum never made it. But its close relative, the sports drink, is booming. And when you talk sports drinks, you're talking Gatorade. It dominates its market like only a few other brands do—Band-Aids in adhesive bandages, Q-tips in cotton swabs.

Gatorade Thirst Quencher—that's the official name—was created in 1965 by University of Florida researchers who were trying to develop a beverage to help the school's football team avoid dehydration in muggy Florida weather. The team's nickname is the Gators and thus the name Gatorade.

And with that they invented a whole new food.

In 1967 the university sold the marketing rights of Gatorade to the Stokely–Van Camp grocery products firm. When Quaker Oats bought Stokely in 1983, industry analysts said Quaker bought Stokely only to get Gatorade.

Even with half a billion dollars in annual sales, the sports drink market is dwarfed by bottled water and orange juice. Let's not even mention soft drinks.

Still, a lot of folks are downing Gatorade, but what is it? Like its competitors—Kool-Aid and Hawaiian Punch—it's sugar water. But it has about half the sugar of Kool-Aid. And it also contains potassium and sodium citrate, minerals to replace the ones lost through perspiration.

That trademark murky green color? It's artificial coloring.

3-PACK HAWAIIAN PUNCH DRINK BOXES— $1.15

Hawaiian Punch was not invented in Hawaii. Nor was it invented by Hawaiians. It was invented in 1936 by a couple of southern Californians, A. W. Leo and Tom Yates.

It actually began as a soda fountain syrup. Mixed with water it was a drink, but it could also be used as an ice-cream topping. By 1944 department stores were selling it in their gourmet food sections, so Leo began bottling it for consumers. That's when he named it Hawaiian Punch. At first it was available only as a syrup, in pint and quart bottles, and Hawaiian Punch lovers had to mix it with water at home. Soon Leo brought it out in a pre-

---

FORKLORE #18

AND IT GETS RID OF WAXY, YELLOW BUILDUP

Lemon Pledge furniture polish contains more lemon than Country Time Lemonade.

---

mixed forty-six-ounce bottle. The drink was available only on the west coast until 1950. It owes a large part of its national popularity to its late-fifties TV commercials with a guy in a Hawaiian shirt offering his friend a Hawaiian Punch and giving him a sock in the puss.

Hawaiian Punch was purchased by cigarette maker R. J. Reynolds in 1963, Reynolds's first venture outside the tobacco field.

Pastas, Macaroni and Cheese

When Judy turns into this aisle, the Muzak that has been playing during our shopping trip is interrupted by a soft voice on the PA system informing her that Kraft Macaroni & Cheese Dinners are an unadvertised special today and the cashier has a twenty-five-cent coupon.

Actually it's not Muzak, it just has that pop-music-run-through-an-Osterizer sound that Muzak has. It's actually a special satellite radio network just for grocery stores. There are several of them in operation and manufacturers like them. The advertising is cheap; it costs a manufacturer about forty dollars per month per store for a thirty-second commercial that runs every hour, twelve hours a day, for thirty days. And it's effective. Studies have shown that a product advertised over the in-store music channel will jump off the shelves with an average increase of 15 percent in sales.

As a matter of fact, we're heading for the macaroni and cheese right now.

7.25 OZ. BOX KRAFT MACARONI & CHEESE DINNER—$0.59

This just may be the first convenience food: Everything in one package, easy to cook, easy to serve. It predates the TV dinner by almost two decades. Kraft introduced it to supermarket shelves in 1936. Actually Kraft, with all its high-paid marketing executives and high-paid food technologists, didn't come up with the idea. That came from one of Kraft's traveling salesmen, Harry Weishaar of St. Louis. At the time, Kraft sold a two-ounce package of grated cheese that could be used on baked dishes, in soups, and in sauces. Grocers were encouraged to run combination sales: soup and cheese at a special price.

Weishaar knew the perfect use for this cheese packet: macaroni and cheese. He would attach the packets to macaroni boxes with rubber bands. Word got back to headquarters and soon the high-paid development staff had come up with the Kraft Dinner, a blue cardboard box with two ounces of cheese, six ounces of macaroni, and instructions on how to "make a meal for 4 in 9 minutes." For only nineteen cents.

It was an immediate hit and Kraft was soon selling 12 million boxes a year. Those sales figures quintupled during World War II. With a little help from the rationing board. Shoppers could get two Kraft Dinners for only one red ration point, if you're old enough to remember what those were.

Today sales peak during Lent and in the early fall. Fall is back-to-school time, when moms appreciate the dinner's quick-fix feature. But Lent? Kraft wondered about that, too. Researchers found no religious preference in sales patterns. What they did find was that spring was when much of the product's promotional dollars were being spent. It was the advertising.

The 1990 model of the Kraft Macaroni & Cheese Dinner is

pretty much the same as the 1936 model. It now takes only seven minutes to cook, instead of the original nine. But that was changed in the thirties. There's a little less cheese: 1 1/4 ounces to the original 2 ounces. The cheese is different, but Kraft says it's improved. It's now spray-dried to get the moisture out. The original was drum-dried and it didn't wring every last bit of moisture out. But the spray-drying process was introduced in 1960. The cheese is artificially colored, but it was back then, too. The package looks pretty much the same as it did in ads back then.

Oh, wait, here's something different: microwave instructions. To cook Kraft Macaroni & Cheese Dinner in a microwave takes longer than stove-top cooking.

I think I like this product even better.

One of the boxes of Kraft Macaroni & Cheese Dinner has the side torn off. This is becoming a more common sight in the supermarket. Shoppers discover after they have pitched a product box into the trash that they needed that UPC symbol for a rebate or a special mail-in offer. The simplest way to replace it is to tear one off a full box in the store.

FORKLORE #19

THEY HAD MACARONI AND CHEESE THEN?

Every good boy knows the song: "Yankee Doodle went to town/A-riding on a pony/Stuck a feather in his hat/And called it macaroni."

But lest you worry that George Washington was trying to chew sticky, starchy macaroni and cheese with those old wooden choppers of his, never fear. The macaroni in the song was a slang term for a dandy from the London Macaroni Club. America didn't have macaroni in Colonial times. Thomas Jefferson tried to import the Italian dish, going so far as to ask his emissary to Naples, William Short, to fetch him a macaroni machine. Short erred, returning with a spaghetti machine instead.

Supermarkets don't make a big deal out of it if they spot some-one doing this. Shoplifting is a much bigger problem. Super-markets lose an estimated $5 billion a year to thieves.

The most-stolen items from supermarkets are cigarettes, health and beauty aids, meat, seafood, and batteries.

Stores are fighting back with elaborate security and closed cir-cuit camera systems. Cameras are located in tiny slots in the ceiling and behind two-way mirrors along walls. In many su-permarkets the only area not covered by a security camera is the bathroom.

And not all of the theft comes from shoplifters. A significant portion is an inside job: a stock boy slipping merchandise in his pocket, a cashier failing to ring up an item in a friend's order. Look straight up the next time you are at the checkout. You should be able to spot the eye-in-the-sky, a camera that can zoom in close enough to read the numbers on the cash register. Most supermarket computer systems are hooked up so that if a cashier is suspected of theft, the tape on his or her register is printed out in the security office. The security officer can zoom in and watch to make sure the items going into the cash register match with the items going into the bag. If a cashier tries to pass some-thing around the scanner, the security officer can spot it on the camera and note on the tape that it was not rung up. Once such crimes were cause for dismissal. Now they are cause for prose-cution.

10 OZ. PACKAGE TOP SHELF ITALIAN-STYLE LASAGNA—$1.79

Here's another food innovation with a terrible name: shelf-stable entrées in retortable tubs. It's a good thing the food com-panies only use names like that in trade conversation.

Shelf stable means the food can be stored on a regular store shelf with no refrigeration. *Retortable* doesn't come from *retort*, meaning witty reply, but from *retort*, a closed lab vessel used for distillation.

When I first saw Top Shelf dinners, I assumed they were some sort of freeze-dried product: "powdered lasagna like the astronauts eat!" Nothing like it. I was surprised the first time I opened one and found it was moist inside. The food is cooked and then vacuum-packed in a closed vessel, a retort.

But a can by any other name is a shelf-stable entrée. And that's really what shelf-stable entrées are: expensive canned goods. Some, like Lunch Bucket, even look like cans. The processing is almost identical to canning. The containers are filled and sealed in a vacuum chamber. Then they are sterilized at 250° F. in a pressure cooker. The food is "shelf stable" because—like a canned good—there is no oxygen inside. But unlike canned goods, many shelf-stable dishes are crunchy, not mushy. That's because they are pressure-cooked about half the time of a canned food. The manufacturers say that makes them crisper and fresher.

Shelf-stable foods were introduced to the supermarket shelves in 1987 by George Hormel, the meat packer. Hormel spent around $25 million to launch its Top Shelf line. Part of the start-up costs were to educate consumers, let them know that a shelf-stable entrée wasn't "powdered lasagna like the astronauts eat!" I must have missed those commercials. Within months General Foods had its own shelf-stable line, Impromptu, and Kraft had its line, La Carte. And Hormel wasn't unhappy about the competition. It meant the cost of consumer education would be shared.

In just a few years shelf-stable entrées have become a $200 million market. But let's not forget that Alpo is also shelf stable.

One secret to shelf-stable products is the package, a tray made of thin layers of polyurethane and aluminum foil. Continental Can worked with Hormel developing the product and the packaging concurrently. Shelf-stable products stay fresh without refrigeration for more than eighteen months, but the dish is so thin that the food cooks in two minutes.

And it doesn't taste too bad . . . for something with a name like shelf-stable entrée in a retortable tub.

4 OZ. CAN STAR-KIST CHUNK LIGHT TUNA IN OIL—$1.19

In the early sixties the most famous animal on TV wasn't a talking horse. Mr. Ed was second. It was a talking tuna named Charlie, the commercial spokesfish for Star-Kist Tuna. Charlie was perpetually disappointed because he wasn't good enough for Star-Kist. Actually Starkist was fudging a bit. They wouldn't have tossed Charlie back because he wasn't prime tuna. No, they would have shifted him over to their cat food factory for use in 9-Lives Cat Food. Of if Charlie wasn't even good enough for cat food, he would have ended up as poultry or livestock feed.

The only way they would have tossed him back was if he had turned out to be a dolphin. Starkist was the first major tuna company to refuse to purchase any tuna caught with dolphins. Other tuna companies quickly followed Starkist's lead.

The primary tuna for Star-Kist canned tuna is the albacore tuna, which spawn in the central Pacific, migrate to the west coast of the United States, and one or two years later are found off the coast of Japan before returning to their spawning grounds. That's a trip of about 15,000 miles, which means that by the time they reach the Star-Kist cannery they are out of warranty.

Frozen Foods

■▲■▲■▲■▲■▲■▲■▲■▲■▲■▲■▲■▲■▲■▲■▲■

I married a microwave. My wife had a microwave when we got married. That was back when pregnant women were advised to stand clear when the microwave door opened and people with pacemakers were advised to move to other neighborhoods.

Our first microwave was an enormous appliance—it took up almost the entire counter next to the sink—and it was almost all working parts. The actual oven was just a small cubicle. The rest of the innards were filled with motors and gears and I-don't-know-what-alls. It looked like a nuclear power plant in a box.

It was called the Amana Radar Range, a high-tech name if I ever heard one. At that time, 1977, only 10 percent of American homes had microwave ovens. The microwave oven took off in the eighties. By 1990, 76 percent of homes had one.

One look at the frozen foods section and you could figure that out. Virtually every frozen food has microwave instructions next to regular oven instructions. And many of the foods have microwave in their names. "Microwave" has replaced "Boil in Bag" and "Heat and Serve" in the names of frozen food products.

The old oven has been left behind. In 1990 a conventional oven was used to prepare meals only 15 percent of the time. And about half of those meals (47 percent) were cooked on the oven's burners and not in the oven itself. For many homes the oven is becoming a decorative hunk of enamel.

The microwave oven is the first advance in cooking since the discovery of fire some 27,000 years ago. It has its roots in the winds of war, the real war, not the book. In 1940 British scientists, led by Sir John Randall and Dr. H.A.H. Boot at Birmingham University, developed microwave radar to detect nighttime attacks by Hitler's Luftwaffe. But no one had time then to worry about inventing microwave popcorn or microwave French fries. That was left to us Americans.

Dr. Percy Spencer, an engineer at Raytheon, was working over a magnetron, the tube that produces microwaves, in his lab in 1946 when he got the munchies. He reached in his pocket for a candy bar he had secreted in there, only to discover a gooey, messy glob. The candy was melted. He knew he hadn't felt any heat, so he wondered if perhaps the magnetron had somehow created heat inside the candy bar. He laid some popcorn near the tube and—presto—the lab smelled like a movie theater. The next day he toted a dozen eggs to work and continued his experiments. He cut a hole in a pot and focused the magnetron's waves through the hole. A curious colleague who had wandered in was greeted with an exploding egg in the face. Dr. Spencer immediately knew he had hit on a new form of cooking because the egg would have had to cook from the inside out for the shell to burst like that.

He told his bosses about his Mr. Wizard–meets–Betty Crocker experiments and they set out to develop a commercial oven. The result was the first Radar Range, a gargantuan stove with so many vacuum tubes and cooling fans that it looked like a prop for *Captain Midnight*. Only a couple of restaurants bought this oversize camp fire, so it was back to the drawing board. In 1952

Tappan brought out the first commercially available microwave oven. This monstrosity was smaller than Raytheon's monstrosity and it had two cooking speeds and a twenty-one-minute timer. The problem was the price, $1,295, about the same as a small home. Tappan didn't sell very many either, but they generated a lot of excitement at home shows.

The microwave boom—and I don't like using that word *boom* in connection with radiation—really didn't start until 1967 when Amana introduced the compact microwave oven.

Still it took twenty years for microwave ovens to take off. In 1986 there were 278 new microwave products introduced. Two years later that number had more than tripled to 962.

The future is even brighter—and I don't like using that word *brighter* in connection with radiation—as an entire generation grows up microwave-literate. Eighty percent of homes with children aged six to fourteen own a microwave! And most of those kids prepare food by themselves at least once a week.

Despite all the microwave entrées on the grocery shelves, cooking dinner isn't the primary use for a microwave oven. According to a 1990 Roper Organization study, the number-one use for the microwave is to reheat leftovers; 91 percent of microwave oven owners said they used it for that purpose. Second is fixing side dishes and snacks, according to 82 percent of the respondents. Preparing frozen food entrées ranks third, tied with defrosting frozen meat.

Our microwave is like our television: It's magic. I don't know how it works and I'm not sure I want to. But here goes anyway:

Microwaves are a form of radiation similar to light. They are fields of electrical and magnetic energy traveling together in waves. Microwaves can be reflected or absorbed depending on the material they are beamed at. Metal reflects; food absorbs. That's why you can cook food with microwaves, but you'll cook the microwave oven if the food is in an aluminum pan. There's a thing called a magnetron in our microwave oven that generates microwaves. The waves are dispersed around the oven's interior and are reflected off the oven walls.

The microwave oven doesn't "cook" the food, it heats the water content of the food and that cooks the food. Almost all food has some water molecules in it. The microwaves are absorbed by water in the food. The water is agitated until it generates heat and that heat cooks the food. The microwaves don't produce heat unless they are absorbed. There's no water in the plate that holds the food, so the plate doesn't get cooked. If the plate is hot to the touch, it is from the heat of the food. So microwaved food is cooked from the inside out. A conventional oven heats the air and this heats the food, from the outside in.

Microwave food sales of all types—frozen and shelf stable—topped $2 billion in 1989. We're making our contribution today.

16 OZ. PACKAGE BIRDS EYE GREEN PEAS—$1.28

Birds Eye is a strange name for a frozen food. Birds Eye?

Today no right-thinking company would put out a food under the brand name Birds Eye. It just doesn't have the right connotation for food. You wouldn't put out a food with the brand name Birds Feet, would you?

FORKLORE #20

WHAT'S IN A NAME?

In 1988 the Adolph Coors Company, brewers of beer, decided they needed to attract younger beer drinkers to their flagship brand, Coors Banquet Beer. How to do it? They brainstormed and talked to consultants and experts and finally arrived at the solution. They'd change the name. Not the ingredients formula. Not the brewing process. Just the name.

And thus was born Coors Original Draft.

Legions of longtime Coors drinkers abandoned the beer because they thought it had changed, too.

Sufficiently humbled, Coors dropped the new label and new name.

But Birds Eye was founded back before that was a consideration. The name comes from Clarence Birdseye, an American businessman who invented the process for freezing foods in packages small enough for retail. Bob Birdseye, which—thankfully—is what he was called, discovered the miracle of freezing food while hunting in Labrador in 1915. The fish and caribou meat that accidentally froze in the dry Arctic air were tender and fresh when thawed and cooked later. He came back home and fiddled around until he figured out how to duplicate that Arctic process commercially. It took a while. In 1923 he founded Birdseye Seafoods. He expanded his product line in 1926 to include other frozen foods. And in 1929 he took the money and ran: He sold out to the rapidly expanding General Foods.

SWANSON HUNGRY-MAN CHOPPED BEEF STEAK DINNER—$3.08

Tomorrow we're taking our youngest son to a basketball game, so Judy is trying to pick out a TV dinner for our other son to fix for himself. She debates between a Banquet Beans and Franks Frozen Dinner and a Swanson Hungry-Man Chopped Beef Steak Dinner, settling on the Swanson. In making her choice, she ignores the taste (our oldest son prefers Banquet) or the packaging (Banquet recently redesigned its package so it is more eye-catching with—wow!—color-coordinated earth tones) and goes strictly with price: Her coupon makes the Swanson dinner fifty cents cheaper.

Her choice goes against recent trends. Price is no longer the main determinant of brand selection: Almost half of all shoppers don't even look at the price of an item before putting it in the cart.

Judy's choice is also an historic one. Swanson was the inventor—albeit a reluctant one—of the TV dinner. Swanson Foods had been selling frozen chicken pot pies, an earlier convenience food, for two years when in 1953 Carl Swanson's sons suggested he offer a complete frozen dinner. He resisted until one M. Crawford Pollock in the Swanson marketing department told him frozen dinners would be the perfect meals for families to eat while watching TV. It would not be the first time that the marketing department of a food company would guide the production department.

So in January 1954 Swanson's TV dinners went on sale.

The first one—turkey and gravy on corn bread with peas and yams on the side—featured a simulated wood-grain TV set on the box. The first TV dinners were three- or four-course frozen meals sold in their own heat-and-serve aluminum trays. They were convenient and cheap and wildly popular: a frozen concoction that would go straight from freezer to oven with no fuss, no muss. The TV dinner and its progeny played a major role in shifting the emphasis in the food industry from taste to convenience.

Business boomed and soon Banquet and Armour joined the competition. In 1990 Americans spent more than $1 billion a year on TV dinners. What they get is a far cry from Carl Swanson's simple frozen meal. Today there are fancy TV dinners with names like Le Menu and Lean Cuisine and Dinner Classics. Many resemble those early TV dinners but with a few additions to protect freshness (BHA), enhance flavor (MSG), retain moisture (sodium tripolyphosphate), and retain color (disodium dihydrogen pyrophosphate).

But the TV dinner market has changed in the last decade and Swanson has failed to change with it. Swanson has no upscale Swanson dinner, no low-cal Swanson dinner. And it was the last to offer a dinner in a microwaveable tray.

LOONEY TUNES ROAD RUNNER CHICKEN SANDWICH MICROWAVE MEAL—$2.48

Two kids, two different frozen dinners. It's the story of our life.

Our youngest doesn't like Swanson or Banquet or Stouffer's. He prefers Looney Tunes Microwave Meals, a new product from

Tyson foods, the chicken people. They come in three flavors: Yosemite Sam Barbecued Chicken, Bugs Bunny Chicken Chunks, and Daffy Duck Spaghetti and Meatballs. They are aimed squarely at the kids with colorful cartoon-character boxes, a game-pack included inside each, and kid-friendly cooking directions. Pretty soon kids won't need moms. Just microwave ovens and a credit card.

In fact, kids' microwaveable meals is the hottest new idea. One brand, My Own Meals, is even being sold in Toys "R" Us stores.

WEIGHT WATCHERS VEAL PATTY PARMIGIANI ZUCCHINI IN TOMATO SAUCE—$2.87

Weight Watchers is an example of an unrelated business that was able to transfer its magic name to the food industry. The Reggie Candy Bar, named for baseball player Reggie Jackson, is an example of one that couldn't.

Weight Watchers was founded in 1963 at a gathering of overweight women in Queens, New York. It was essentially a support group to help the women stick to their diets. In the early seventies founders Jean Nidetch and Felice and Albert Lippert branched out into the grocery store, selling frozen entrées that matched the Weight Watchers philosophy. In 1978, when the giant H. J. Heinz Company purchased Weight Watchers for $100 million, its frozen entrées were mostly fish-and-vegetable combinations. Heinz food scientists added to the line, spicing it up with more attractive and diverse foods. There's only so much you can do with frozen fish.

3-PACK MICRO-MAGIC MICROWAVE FRENCH FRIES—$1.47

This is the nineties. Most of the French fries consumed at home are of the frozen variety. And since most of the French fries in fast-food restaurants are also of the frozen variety, we're used to the taste.

Frozen French fries are an American invention. But French fries really were invented in France, in the 1700s. Thomas Jefferson brought them back to this country after his stint as ambassador to France ended in 1789.

Frozen French fries were first sold on February 16, 1946, in Macy's in New York City. They were from Maxson Food Systems of Long Island City, a company that is no longer in business. Potato consumption had declined steadily from 1910 and there was hope in the spud industry that easy-to-fix frozen French fries might turn the tide. They didn't. In fact, potato consumption didn't turn around until 1962 and we all know the

---

FORKLORE #21

WAIT TILL THE MIDNIGHT HOUR

Ever wonder what goes on in your local supermarket while you sleep?

Well, stock clerks are mopping floors, dusting merchandise, restocking shelves, rearranging existing stock, and turkey bowling.

Turkey bowling?!

In those early A.M. off-moments many stock clerks will gather together in the back of the store for a little friendly line of turkey bowling. A frozen Butterball turkey serves as the ball and ten 2-liter soft drink bottles are arranged pin-style.

Derrick Johnson of Newport Beach, California, is the commissioner of the PBA—the Poultry Bowling Association. He told *USA Today:* "We're taking turkeys where turkeys have never gone. This is the only sporting event in the world where, after you get done with the equipment, you can eat it."

Johnson was fired for his candid talk. He said he got phone calls from sympathetic grocery store clerks from all over the country, many of whom told him about their late-night escapades in the aisles of America's supermarkets: cantaloupe basketball, Spam slam, even live lobster races.

reason for that: McDonald's and other fast-food joints.

Those first frozen fries were prepared in much the same way that today's frozen French fries are made. The potatoes were briefly fried for one to two minutes before freezing. Until the eighties most frozen French fries were baked in the oven. The microwave has changed that. But the microwave doesn't brown foods and since no one wanted to eat snow-white French fries, manufacturers had to come up with a way to brown the fries. Amazingly enough, they didn't turn to artificial coloring. Instead they turned to the packaging industry, which invented something called a "heat susceptor." It's a thin piece of metalized plastic that's built into a cardboard carton and it has many uses, from browning frozen French fries to crisping the crust of a frozen pizza to popping those pesky last few popcorn kernels

FORKLORE #22

GREAT DATES IN FOOD HISTORY

Here are the dates when some of America's favorite foods were introduced:

1896—Tootsie Roll
1897—Jell-O
1897—Grape-Nuts
1907—Hershey's Kisses
1912—Goo Goo Clusters
1912—Life Savers
1912—Oreos
1914—Clark Bar
1914—Mary Jane
1916—All-Bran
1921—Mounds
1923—Milky Way
1923—Reese's Peanut Butter Cup
1927—Kool-Aid
1928—Rice Krispies
1930—Birds Eye Frozen Foods
1930—Snickers

1932—3 Musketeers
1934—Ritz Crackers
1941—Cheerios
1944—Hawaiian Punch
1946—Minute Rice
1947—Almond Joy
1950—Sugar Corn Pops
1952—Kellogg's Sugar Frosted Flakes
1953—Sugar Smacks
1956—Duncan Hines Brownie Mix
1956—Jif Peanut Butter
1958—Tang
1965—Shake 'n Bake
1966—Cool Whip
1968—Pringles
1976—Country Time Lemonade
1981—Butter Flavor Crisco

in microwave popcorn. The susceptors absorb some of the microwave energy and become like a tiny hot plate inside the box, emitting heat and browning the fries.

If you want to see the susceptor in your French fry box, hold the package up to the light.

JENO'S CRISP 'N TASTY FROZEN PEPPERONI PIZZA—$2.39

Who eats frozen pizza?

The typical heavy user is a high school graduate living in the southern or west-central United States, nonmetropolitan, aged thirty-five to forty-four, a parent with five or more people in the household.

In other words, someone with kids who lives outside Domino's delivery area.

We qualify on both counts. Tonight it's the kids factor. We have a baby-sitter coming tomorrow night and we need something she can fix and our son will eat. The brand of frozen pizza we buy is dependent on one factor and one factor alone: What will our nine-year-old eat? We've tried various brands over the years. His current choice is Jeno's.

Nationally Jeno's is the third most popular frozen pizza.

Frozen pizza has only one competitor: home-delivered restaurant pizza. Homemade pizza has never taken off in this country. That's because you need a professional baker's oven to give pizza that trademark crispy crust, chewy center, and bubbly cheese top.

And no one will disagree that restaurant pizza is the best. Perhaps that's why Domino's alone sells more pizza than all the frozen pizza companies combined. Frozen pizza sales totaled a mere $659 million in 1990. Domino's sales were $2.7 billion that same year.

Frozen Desserts

▆▗▆▗▆▗▆▗▆▗▆▗▆▗▆▗▆▗▆▗▆▗▆▗▆▗▆▗▆▗

Humans have always tried to figure out ways to cool their food and drink. Alexander the Great filled trenches with snow and iced down wine kegs in them. He passed out the chilled beverage to his soldiers on the eve of battle, giving new meaning to the expression "fighting a hangover." The Roman emperor Nero sent his slaves to the mountains to retrieve snow to cool his wine. Some tribes of North American Indians would freeze meat and fish in the winter snows. British philosopher Francis Bacon even applied his famous inductive reasoning powers to the matter. He stuffed a duck with snow to see if it would preserve and sterilize the bird but didn't live to see the result of his experiment. He caught cold and died.

For centuries the problem faced by those wanting to refrigerate food was finding ice. The first ice used for food cooling and preservation in this country was harvested from lakes and ponds. Which meant you had plenty of ice in the wintertime, when you didn't need it, and virtually none in the summer, when you did.

That didn't prevent an industry from springing up around ice harvesting. Ice was stored in caves and especially constructed family icehouses—Washington, Jefferson, and Monroe all had them.

The first refrigerator—we'd call it an icebox today—was invented in the early years of the nineteenth century by Maryland farmer Thomas Moore. His design, inspired by a water cooler diagram in the book *Hepplewhite's Cabinet Maker and Upholsterer's Guide*, which had been published in London in 1789, involved situating a tin vessel inside an oval of wood, filling the space between the tin and the wood with ice, and covering the contraption with rabbit skins for insulation. It was a bit cumbersome to get in and out of, so Moore redesigned it, fitting a wooden box inside a wooden box, insulating the walls with charcoal and ashes, and situating the ice in a vessel on top. He patented this refrigerator design on January 27, 1803. Later that same year he published a treatise on the future of his invention, "An Essay on the Most Eligible Construction of Ice-Houses, Also a Description of the Newly Invented Machine Called the Refrigerator," in which he predicted that in the future the refrigerator would be used to carry dairy products and meats to market, to keep foods fresh in the store, and to preserve foods at home until mealtime. Little did he know.

The first home iceboxes were being sold as early as 1830. Home iceboxes held anywhere from 50 to 125 pounds of ice. During warm weather the iceman made his rounds every day. A block of ice would last about a day, if the icebox's doors weren't opened too frequently. Some practical people tried to extend the life of the ice by wrapping it in a blanket. It worked. Of course, the food spoiled. But the ice lasted longer.

The first artificial ice-making machine was invented in 1834 by Jacob Perkins of Newburyport, Massachusetts. His machine evaporated sulfuric ether with an air pump, an early form of compression refrigeration. In 1853 A. C. Twinning of New Haven, Connecticut, patented a machine that would make a ton of ice a day.

The first electric refrigerator for the home, the Domestic Electric Refrigerator, was sold in Chicago in 1913. In 1914 Kelvinator introduced its electric refrigerator. Soon the field was flooded with manufacturers. By the twenties the ice-harvesting business had gone the way of wooden-teeth makers.

8 OZ. TUB COOL WHIP—$1.08

Corn flakes, graham crackers, and Snickers Bars were all invented long before I was born. But I can remember when they brought out Cool Whip. My mother said it would never replace whipped cream in our household.

She now has a freezer full of Cool Whip and a cupboard full of old Cool Whip canisters.

What a great product! It never goes away. After you eat the fluffy, almost tasteless white stuff, you have a lifetime supply of storage canisters. Right now my refrigerator has an old Cool Whip tub full of green beans, an old Cool Whip tub full of Chef Boyardee ABC's & 123's, and an old Cool Whip tub full of something I don't immediately recognize.

We have to keep buying the stuff just for the reusable plastic tubs. Either that or go to a Tupperware party.

Cool Whip came on the market in 1967. It was an early success of marketing research. General Foods reseachers asked housewives about products they'd like to see and, according to an article in a 1967 issue of the in-house newsletter *General Foods News*, the housewives told them they wanted "a processed dessert topping that came already whipped; one that, unlike others on the market, required no preparation, but could be spooned right onto a dessert [and] taste as good as whipped cream." That sounds like a housewife talking, doesn't it?

That wasn't all they wanted: They also wanted "longer refrigerator storage, lower cost, less calories." Pretty demanding, those consumers, but General Foods' Technical Center people were up to the challenge. They whipped up—so to speak—the product in six months.

Birds Eye, General Foods' frozen foods division, test-marketed Cool Whip in April 1966 in Seattle and Buffalo; Seattle because it was "a city where usage of fresh whipped cream is high," Buffalo because it was "where processed toppings are quite popular."

Cool Whip was a huge success. By Christmas 1966, 60 percent of the housewives who had sampled Cool Whip had made a repeat purchase.

From its inception Cool Whip was referred to as "non-dairy whipped topping." There was no intent to fool those gullible housewives into thinking it was whipped cream. One look at the price tag and they knew it wasn't. Cool Whip sold for about one-fourth the price of canned whipped cream.

So what is Cool Whip if it isn't whipped cream? Mostly water. Cool Whip is made from water, corn syrup, hydrogenated coconut and palm kernel oils, sugar, sodium caseinate, polysorbate 60 and sorbitan monostearate, natural and artificial flavors, xanthan gum and guar gum, and artificial colors. (It isn't even really white!)

What is all that stuff?

Water, you know about.

Corn syrup is like Karo Syrup. In short, it's sugar.

Hydrogenated coconut and palm kernel oils are processed vegetable oil: sort of like hardened Wesson Oil. I couldn't begin to explain the hydrogenation process—it's called a "physio-chemical modification" combining hydrogen and oil—but I can tell you what it does. It turns vegetable oil liquid into a solid, in the process removing certain fatty acids that turn rancid easily. The hydrogenation process also makes it difficult for the oil to revert back to its liquid form.

Sugar, you know. It's a dirty word now. It's also everywhere. With corn syrup as the second ingredient—meaning it is second in weight—and sugar as number four, that means Cool Whip has a lot of sweeteners.

Sodium caseinate is derived from milk protein. It gives Cool Whip its texture. It's also used in glue, plastic, and paint.

Polysorbate 60 and sorbitan monostearate are emulsifiers; they keep the water and oil from separating. Polysorbate 60 is made from sorbitol, a sugar alcohol, and fatty acids from vegetable or animal oils.

Natural and artificial flavors means you have no idea what this concoction really tastes like.

Xanthan gum and guar gum are derived from plants. They are used here to thicken and to hold the water and oil together.

Artificial colors . . . well, you know what they're all about.

The label touts "only 12 calories per serving" with "per serving" in letters a small ant would have trouble reading. Then there's a picture of Cool Whip swirled on a dish of strawberries. Must be specially bred pygmy strawberries because, according to the side label, one serving is one tablespoon. There are fifty-six servings in the container!

8-COUNT AUNT JEMIMA HOMESTYLE (FROZEN) BLUEBERRY WAFFLES—$1.48

When is a blueberry not a blueberry? When it is a "blueberry bit," the industry term for the blueberry-looking things that are found in frozen waffles and pancakes. The name "blueberry bit" has a wonderful sound to it. But a blueberry bit is not a blueberry. It's a little piece of molded sugar dough artificially flavored to taste like a blueberry and artificially colored to look like a blueberry.

Here's what a blueberry bit really is: "sugar, dextrose, partially hydrogenated cottonseed oil, artificial flavor, salt, citric acid, cellulose gum, enzyme modified soy protein flour, corn syrup solids, silicon dioxide, FD&C blue number 2, malic acid, sodium hexametaphosphate, and FD&C red number 40." Somehow that just doesn't conjure up images of springtime in Maine, does it?

12-COUNT BOX JELL-O CHOCOLATE FLAVOR PUDDING POPS—$2.58

If General Foods had introduced Jell-O Pudding Pops in 1953 instead of 1983, it might have named them something else: ChocoPops or Cream Pops or Dream Pops. But it for sure wouldn't have named them Jell-O Pudding Pops because they don't have Jell-O pudding in them.

Compare the labels:

Jell-O Chocolate Fudge Flavor Pudding & Pie Filling: sugar, cornstarch modified, cocoa processed with alkali, sodium phosphates, artificial flavor, di- and monoglycerides, artificial color, hydrogenated soybean oil with BHA, nonfat milk, hydroxylated soybean lecithin, natural flavor.

Jell-O Chocolate Flavor Pudding Pops: skim milk, sugar, water, nonfat dry milk, hydrogenated coconut and palm kernel oils, cocoa processed with alkali, corn syrup, modified tapioca starch, dextrin, sodium caseinate, salt, artificial flavor, dextrose, sodium stearoyl lactylate and polysorbate 60, microcrystalline cellulose, sorbitan monostearate, xanthan gum, carrageenan, cellulose gum, and guar gum.

They both contain sugar and cocoa and some form of milk. And a lot of chemicals. And if they have a similar taste, we can lay that to the cocoa and the chemicals.

The Jell-O name on this product comes from a new trend in food: flanker products and line extensions. If General Foods had named its frozen bar something new, it would have had to spend millions of dollars in advertising and promotion to create consumer awareness of this new product. If it swipes a recognized brand name from one of its established lines, its new product has instant name recognition and—General Foods hopes—instant sales. It's happening up and down the grocery shelves.

When Procter & Gamble brought out a new line of mouthwash, it didn't call it Fresh Breath. It called it Crest and the new mouthwash automatically inherited the image of the company's well-known toothpaste. The strategy has been so successful that the Crest name is used on thirty-eight different products.

In some cases a brand name line is extended into a new area: Chiquita Bananas begat Chiquita Fruit & Ice Cream Swirl Bars.

Kraft Velveeta Cheese begat Kraft Velveeta Shells & Cheese. In other cases it's merely a side-by-side extension: a new flavor or size of an existing brand.

It's all in the name, the brand name. Brand names are now one of a company's most valuable assets. When Philip Morris paid $13 billion for Kraft, Inc., in 1988, some analysts said it was buying Kraft's cache of well-known brand names.

There were no brand names in the old trading post. Salt was salt. Bullets were bullets. Blankets were blankets.

As shopping shifted to general stores, it was still a while before consumers became acquainted with brand names. The crackers in the cracker barrel were just crackers and if anyone complained, the storekeeper just started buying them from someone else.

The first food to be trademarked was Underwood Deviled Ham in 1843. Brand names were an immediate hit in the grocery stores. Customers wanted consistency and quality and that's what a brand name promised them. Americans—who were descended from many religious sects that believed cleanliness was next to godliness—also liked the hygienic guarantees of canned and packaged brand-name goods.

Today the brand name is almost as important as the product itself. It's the billboard, the first thing a shopper notices, and it needs to tell the shopper a lot in a little time. Many a product failure has been blamed on the name.

Gone are the days when a manufacturer could name his newest product after his baby daughter, as John Markel did when he named his new lollipop the Tootsie Roll after his little girl Tootsie. Now companies employ high-priced identity consultants to come up with names.

David R. Wood, a Chicago identity consultant, has named some 500 products since 1979, including Pepsi's Slice orange drink, Polaroid's Spectra camera, Tylenol's Gelcaps safety-coated headache medicine, and Bristol-Myers's Nuprin pain pill. His fee starts at $30,000. The fee is high because the stakes are high and Wood has a good track record. Launching a new product can cost anywhere from $10 million to $150 million. But the rewards can be even higher. Pringles Potato Crisps topped $100 million a year in sales within a couple of years of its introduction.

It's not easy coming up with a name—consultants use computers, phone books, anything to help—but it's even harder coming up with a name that hasn't already been trademarked. In 1990 more than 50,000 trademarks were registered with the U.S. Patent and Trademark Office, three times the number registered just ten years earlier.

The only section of the supermarket to escape this brand mania is the produce section, where the only brand names of any consequence are Dole, Chiquita, and Sunkist. But rest assured someone is eyeing produce for brand possibilities.

What would happen if we had logo lettuce or trademarked tomatoes? The same thing we have in the meat department, where name brands were unheard of before Frank Perdue Chicken: higher prices.

Somebody has to pay for those trademark searches.

I PINT BEN & JERRY'S CHERRY GARCIA ICE CREAM—$2.75

We are definitely an ice cream–loving country. With 9 percent of the world's population, we consume 45 percent of the world's ice cream. And ice cream has gone uptown in the last two decades. There's still old-fashioned Neapolitan ice cream and vanilla ice milk, the diet ice cream of the sixties, but the growth in the industry is in what they call "superpremium ice cream." Rich, creamy, gooey, tasty ice cream, full of lots of butter and fat. Ice cream that is an indulgence. There's none of the old *Playboy* rationale: "I only buy it for the articles." You don't buy superpremium to satisfy your minimum daily requirement for calcium. You buy it for the taste. Superpremium ice creams contain 20 percent milkfat, twice that of ordinary ice cream.

When it comes to superpremium ice cream there are three names: Häagen-Dazs, a Pillsbury company with a made-up name created to sound Scandinavian; Frusen Gladje, a Kraft company with a made-up name created to sound like Häagen-Dazs; and Ben & Jerry's, a company with a name taken from founders Ben Cohen and Jerry Greenfield.

We've bought Ben & Jerry's ever since they successfully turned back a Häagen-Dazs/Pillsbury plot to keep them off the grocery shelves. Häagen-Dazs told Northeast distributors to drop Ben & Jerry's, or risk losing Häagen-Dazs. After the story hit the newspapers, Häagen-Dazs backed off. But the publicity gave Ben & Jerry's a boost and helped them grow beyond their Vermont base.

We just like their attitude. They bypassed Wall Street when they took the company public, selling their initial public offering themselves; they solved a waste disposal problem by buying pigs to eat ice cream by-products; and they named a flavor Cherry Garcia, after Jerry Garcia, the guitarist for the rock group The Grateful Dead.

Cohen and Greenfield were certainly helped by all their well-publicized shenanigans. But the thing that made Ben & Jerry's Ice Cream a hit was their savvy niche marketing: They found a niche in the ice cream market that Pillsbury and Kraft had overlooked and exploited it. In this case it was in flavors. While Häagen-Dazs and Frusen Gladje were selling boring flavors—coffee ice cream was probably the most exciting—Ben & Jerry's was offering such flavors as New York Super Fudge Chunk—chocolate ice cream swirled with white- and dark-chocolate chunks, almonds, pecans, and walnuts—and Cherry Garcia.

Actually the Häagen-Dazs power play was out of character for the brand. Häagen-Dazs was created in 1960 as a reaction to the big guys. The large ice cream companies were involved in a price war that forced many of the little guys out of the business.

Ruebin Mattus, a Polish immigrant who had been selling ice cream in the Bronx for more than sixty years, thought maybe people were tired of the cheap air-fluffed ice cream and would be willing to pay for a top-of-the-line brand. His new ice cream didn't have nearly as much air pumped into it as other brands. And it was expensive. It cost seventy-five cents a pint at a time when other brands were selling for fifty-three cents a pint.

What gave it a jump start was the name: Häagen-Dazs. His wife, Rose, made it up. She was trying to create a name that sounded important. The crowning touch was the umlaut over the first *a*.

If Mattus had still been in charge, Häagen-Dazs would probably not have pulled its power play. But Pillsbury bought the company from him in 1983 for $80 million.

Judy and I have no excuse for buying this rich, expensive ice cream. Friends aren't coming over; Judy isn't planning a bridge party. It's just a Friday night indulgence, a reward for surviving another week. We'll go home and have a dish tonight. And by this time tomorrow night all the ice cream will be gone.

When I think of ice cream, I think of a summertime treat, something you eat when the sun is beating down on your back. But not anymore. Now ice cream is a nighttime snack. Most ice cream is consumed between nine and eleven P.M.

The Statens are what the supermarket researchers call "heavy ice cream users" and, yes, you can interpret that phrase in several ways. Some 16 million households—about 27 percent of ice cream buyers—fit in that classification. It means we buy a half gallon or more a week, accounting for 70 percent of all ice cream purchases.

The heaviest users of ice cream are families with children—that's us.

HALF GALLON MINUTE MAID 100 PERCENT PURE ORANGE JUICE—$2.38

I've been hinting lately that perhaps what we need most is a juicer. The bottled orange juice, the frozen orange juice, none of it tastes as good as fresh-squeezed.

That's what I like about our Kroger's grocery store: They have a juicer and sell fresh-squeezed orange juice. What I don't like is that it's six miles away.

The big orange juice processors know this; they know that frozen orange juice doesn't taste as good as fresh. And they're working on it. The problem is complicated by the fact that federal regulations bar the addition of anything to orange juice if it is sold as "100 percent orange juice." Only the vapors created during processing may be captured and added later.

In their quest to make processed orange juice taste like fresh-squeezed orange juice, scientists are using the latest high-tech lab instruments to pin down the chemical differences between the two. In a 1990 report to the American Chemical Society, USDA scientist Philip Shaw said that the difference between fresh and processed orange juice can be attributed to the quantities of about twenty chemicals that make up what we think of as orange flavor.

Scientists call it flavor. We call it taste. And it is probably the determining factor in every food purchase. It's why applies outsell rutabagas, French fries outsell au gratin potatoes, and Kellogg's Frosted Flakes outsells Kellogg's Corn Flakes. We like the taste better.

In the sixties the Institute of Nutrition of Central America and Panama created a cheap high-protein food to prevent malnutrition in underdeveloped countries in that area. This concoction of cereal, oilseed meal, and a variety of vegetables was a miracle food all right, providing a balanced diet for those who couldn't afford one. The problem was the people in these countries wouldn't eat it. They didn't like the taste.

Looks are important. Aroma is important. But taste is the most important thing in food.

To get the taste of the real thing food manufacturers work with flavor chemists, who can either create a flavor or duplicate a flavor. If a company calls for, say, strawberry, the flavor company might send them its basic stock strawberry flavor and see if they like it. They might like it but say it's too "green" or too "cooked." That's when the flavor chemist goes to work, adding ingredients—mixing potions like a sorcerer—until the flavor is just right.

When a flavor chemist is trying to duplicate a flavor, he'll live with it for several days, carrying a blotter with the smell around in his car, sleeping with it on his pillow. Then he goes to work, scribbling out a list of all the ingredients he could use to make the flavor, pulling them off the stock shelf, and experimenting. Flavorists usually start with a sugared water base and add ingredients until it tastes right.

One of the most legendary cases in flavor industry history was the case of the tomato flavor for a new pasta sauce. The food chemist kept sending it over to the company and they kept sending it back, saying it wasn't quite right. Finally the chemist had them fix some homemade spaghetti so she could see what it was in the sauce they were trying to duplicate. She took one bite and got it. They wanted the tinny taste of canned tomatoes!

The Food and Drug Administration recognizes two kinds of flavors:

Natural flavor (or natural flavoring) is "the essential oil, oleoresin, essence or extractive, protein hydrolysate, distillate or any product of roasting, heating or enzymolysis which contains the flavoring constituents derived from a spice, fruit or fruit juice, vegetable or vegetable juice, edible year, herb, bark, bud, root, leaf, of similar plant material, meat, seafood, poultry, eggs, dairy products, or fermentation products thereof, whose significant function in food is flavoring rather than nutritional."

Artificial flavor (or artificial flavoring) is "any substance, the function of which is to impart flavor which is not derived from the sources indicated above."

An essential oil can be obtained by hand or mechanical pressing like squeezing the juice from an orange or by distilling. These are the truest flavors, although distilling may cause some minor flavor changes. Water-soluble chemicals are lost and heat damage occurs during distillation.

Oleoresin flavors are obtained by percolating: like water percolating through coffee grounds.

Essence or extractive flavors come from a concentrate from the natural source.

Protein hydrolysate are from the flavor fragments of protein obtained by chemical splitting.

Roasting flavors are the flavors produced by heating a flavor source.

And enzymolysis flavors are created by enzyme reactions.

Why do we need extra flavors? Maybe there aren't enough cherries in a cherry pie to flavor the pie fully. Cherries differ in sweetness and flavor, so adding flavor will make the product more consistent. If the manufacturer adds additional cherry flavor, derived from the cherry itself, the label will say "cherry flavored pie" or "natural-cherry-flavored pie." If another flavor is added, it will say "cherry-flavored pie with other natural flavor" or "natural-cherry-flavored pie with other natural flavor."

If an artificial flavor is added, "artificial" replaces "natural" in the name. You can call your product "vanilla pudding" if you use vanilla extract. Otherwise it's "artificially flavored vanilla pudding." And only your taste buds know for sure, most of the time.

37 OZ. MRS. SMITH'S NATURAL JUICE APPLE PIE—$4.98

In their influential 1933 book, *100,000,000 Guinea Pigs*, authors Arthur Kallet and F. J. Schlink asked the question, "Who but the starving would buy a pie labeled thus: 'cornstarch-filled, glucose-sweetened pie made with sub-standard canned pineapple, artificial (citric acid) lemon flavor, and artificial coal tar color'?"

The answer today is: We will.

The label on our pie isn't as blunt as Kallet and Schlink's imaginary label, but it's nowhere near as appetizing as the name, Mrs. Smith's Natural Juice Apple Pie.

Here's what's inside our pie: apples, wheat flour, sugar, margarine (partially hydrogenated soybean oil, soybean oil, water, salt, nonfat dry milk, lecithin, mono- and diglycerides, may contain sodium benzoate, citric acid or calcium disodium EDTA/preservative, artificial color and flavor, vitamin A palmitate), vegetable shortening (partially hydrogenated soybean oil), water, modified food starch, salt, dextrose, whey, spices, baking soda, malic acid, sodium busulfite (preservative).

So how can a pie with the word "Natural" in large script letters contain artificial color and flavor, preservatives, and all those other funny-sounding names?

FORKLORE #23

IS IT LIVE OR IS IT MEMOREX?

Everyone knows there's an Orville Redenbacher. Who would make up a name like that? Besides, we've all seen him—and his nerdy grandson—on TV commercials. And everyone knows there is no Betty Crocker. Every time they change her picture, all the newspapers do a big story about the changing face of Betty Crocker.

But what about those other famous food folks, like Mrs. Paul and Mrs. Smith and Mrs. Filbert?

First the crushing news: There is no Chef Boyardee. And there is no Mrs. Paul.

Chef Boyardee's name is a combination of the names of the three founders of the company: Boyd, Art, and Dennis. Mrs. Paul is fictitious, although there was a Mr. Paul, sort of, for a while in the late eighties. Campbell Soup, makers of Mrs. Paul's frozen fish products, featured Mr. Paul in a series of TV commercials. He was played by John L. Kelly, a former narcotics agent who had one of the greatest jobs in history: He worked as a bodyguard to Elvis Presley's bodyguards.

But there is a Sara Lee. And a Ben and Jerry. And there was a Duncan Hines (he died in 1959).

Sara Lee is Sara Lee Lubin Schupf. (Now maybe you understand

continued

why they just call the company Sara Lee.) Her father, Charles Lubin, created the first Sara Lee product, a cheesecake, in 1949 when she was nine and named his fledgling company after her.

Duncan Hines was a traveling printing salesman and amateur restaurant critic who self-published a restaurant guide as a Christmas present for his friends in 1935. It contained reviews of 167 spots in thirty states. The response was so positive that the next year he expanded it to include 490 restaurants and published it in a paperback edition. His *Adventures in Good Eating* guides became an annual publishing event. In 1948 he lent his name to a line of convenience cake mixes manufactured by Procter & Gamble.

There was no real Aunt Jemima, but there was an Aunt Jemima character. The brand name was inspired by an 1889 vaudeville routine performed by blackface comedians Baker and Farrell to a tune called "Aunt Jemima." When the Aunt Jemima company decided to exhibit at the 1893 Chicago World's Fair, they hired Nancy Green, a Chicago restaurant cook, to play auntie at the fair booth. She continued to impersonate Aunt Jemima until her death in 1923.

There was never a Betty Crocker. Betty Crocker was a creation of the Home Services Department of the Washburn Crosby Milling Company of Minneapolis, a forerunner to General Mills. The department needed a name to sign to letters answering housewives' questions, so in 1921 someone created the name Betty Crocker. "Crocker" was in honor of recently retired company director William Crocker. "Betty" was picked just because somebody liked the name Betty. The female employees of the company entered a penmanship contest to see who got to do the Betty Crocker signature. A secretary won and her penmanship is still in evidence on Betty Crocker packages.

The first picture of Betty was created in 1936 by New York artist Neysa McMein, who used features from the various women in the Home Service Department. Betty's portrait was updated in 1955, but instead of getting older, she actually got a younger appearance so the moms of baby boomers could identify with her. Her picture was updated in 1965 and again in 1980.

The 1965–80 Betty has a striking similarity to Marilyn Quayle. I don't know if that tells us something about Betty or something about Marilyn.

If you ever diagramed a sentence on the board in sophomore English class, you already know the answer. "Natural" doesn't modify "Apple Pie." It modifies "Juice." The pie contains natural juice, so the name is correct, if misleading.

Labels have come a long way since 1906 when the first federal food law was passed. But to call them misleading isn't misleading.

For instance, a product can scream "No Preservatives" on the label, but that doesn't mean no chemicals. The government permits any of thirty-three kinds of additives and only two of those are preservatives.

Diet Coke can advertise itself as the one-calorie soft drink. That's true, if you drink the suggested serving size, six ounces, which is half a can. Not that two calories, a full can, is any diet buster. But take Entenmann's fat-free Chocolate Loaf Cake with "only 70 calories per serving." A serving is one ounce, and if you can satisfy your chocolate genes with a one-ounce serving, you are a rare person.

Turkey hot dogs are labeled 80 percent fat free. That means they are 20 percent fat by weight. And, more insidiously, that means they get 80 percent of their calories from fat. Health organizations recommend we get no more than 30 percent of our calories each day from fat.

In 1991 the Food and Drug Administration seized 2,000 cases of Citrus Hill Fresh Choice orange juice because it wasn't fresh; it was made from concentrate. Fresh Choice was the trade name. The FDA also ordered Mazola to remove "No Cholesterol" from its label because it had never contained cholesterol.

The government is beginning to clean up this labeling mess. There will be strict definitions for what is "low fat" and "low cholesterol," plus mandated definitions for serving sizes using common household measures.

Under the new law food makers will no longer be able to hide the fact that a product is half sweeteners by listing sugar, honey, corn syrup, and other sweeteners separately so they are scattered in the ingredients-by-weight list. They must be lumped together.

Lean won't mean whatever the meat packer wants it to mean. "Lean" will mean less than 10.5 grams of fat (no more than 3.5 grams from saturated fat). And "extra lean" will have less than 4.9 grams of fat (less than 1.8 grams of saturated fat).

Currently a label has to include some common name that identifies what the product is. Strict rules apply here. If a drink contains only a small amount of orange juice, it can't be called orange juice, it must be called orange drink or imitation orange juice or orange-juice-flavored drink.

The label must also give the net weight of the product in the container, the name and address of the manufacturer, packer, or distributor, and a list of ingredients, in descending order by weight. If the manufacturer has added nutrients or makes a nutritional claim for the product, then the label must also list certain nutrition information: the amount of protein, vitamins A and C, the three B vitamins—niacin, thiamine, and riboflavin—and the minerals calcium and iron. These are expressed in percentage of the U.S. Recommended Daily Allowance. Anything else is up to the manufacturer.

I don't think too many of us eat apple pie for its nutritional value, but Mrs. Smith's has all that stuff listed, plus the amount of calories, carbohydrates, fats (broken down into monounsaturated, polyunsaturated, and saturated), cholesterol, sodium, and potassium.

If you're curious, I can get 2 percent of my Recommended Daily Allowance of vitamin A by eating a 4.6 ounce serving (one-eighth of the pie). Fifty pieces and I don't have to worry about any more A for the day.

20 OZ. BANQUET READY-TO-BAKE CHERRY PIE—$1.68

Banquet can call this a cherry pie because it meets the federal "standard of identity" for a cherry pie.

That's right, there's even been government intervention in something as American as cherry pie. Apple pie, too. The government has what it calls standards of identity for a large number of foods. It's sort of like a recipe for the product, so that you can be sure when you buy, say, a jar of mayonnaise, you are getting what you expect to get and not some whitish salad dressing.

Cherry pie has been a part of children's nursery rhymes for hundreds of years. But could Billy Boy bake a cherry pie today? He couldn't call it a cherry pie unless it met this government "Standard of Identity for Frozen Cherry Pie," as published in the *Code of Federal Regulations* in 1974:

> 28.1 *Frozen cherry pie; identity; label statement of optional ingredients.*
>
> (a) *Frozen cherry pie (excluding baked and then frozen) is the food prepared by incorporating in a filling, contained in a pastry shell mature, pitted, stemmed cherries that are fresh, frozen and/or canned. The top of the pie may be open or it may be wholly or partly covered with pastry or other suitable topping. Filling, pastry, and other topping components of the food consist of optional ingredients as prescribed by paragraph (b) of this section. The finished food is frozen.*
>
> (b) *The optional ingredients referred to in paragraph (a) of this section consist of suitable substances that are not food additives as defined in section 201 (s) of the Federal Food, Drug, and Cosmetic Act or color additives as defined in section 201 (t) of the act; or if they are food additives or color additives as so defined, they are used in conformity with regulations established pursuant to section 409 or 706 of the act. Ingredients that perform a useful function in the formulation of the filling, pastry, and topping components, when used in amounts reasonably required to accomplish their intended effect, are regarded as suitable except that artificial sweeteners are not suitable ingredients of frozen cherry pie.*
>
> (c) *The name of the food for which a definition and standard of identity is established by this section is frozen cherry pie; however, if the maximum diameter of the food (measured across opposite outside edges of the pastry shell) is not more than 4 inches, the food alternatively may be designated by the name frozen cherry tart. The word "frozen" may be omitted from the name on the label if such omission is not misleading.*
>
> (d)(1) *Each of the optional ingredients used shall be declared*

on the label as required by the applicable sections of part 1 of this chapter.

(2)The label shall not bear any misleading pictorial representation of the cherries in the pie.

28.2 Frozen cherry pie; quality; label statement of substandard quality.

(a)The standard of quality for frozen cherry pie is as follows:

(1)The fruit content of the pie is such that the weight of the washed and drained cherry content is not less than 25 percent of the weight of the pie when determined by the procedure prescribed by paragraph (b) of this section.

(2)Not more than 15 percent by count of the cherries in the pie are blemished with scab, hail injury, discoloration, scar tissue, or other abnormality. A cherry showing skin discoloration (other than scaled) having an aggregate area exceeding that of a circle nine thirty-seconds of an inch in diameter is considered to be blemished. A cherry showing discoloration of any area extending into the fruit tissue is also considered to be blemished.

(b)Compliance with the requirement for the weight of the washed and drained cherry content of the pie, as prescribed by paragraph (a)(1) of this section, is determined by the following procedure:

(1)Select a random sample from a lot:

(i)At least 24 containers if they bear a weight declaration of 16 ounces or less.

FORKLORE #24

IT HAPPENED AT THE WORLD'S FAIR

Necessity was the mother of the invention called the ice cream cone. It happened at the world's fair, the 1904 St. Louis World's Fair. Ernest Hamwi had a concession booth on a fair aisle called Constantinople-on-the-Pike and sold a waffle pastry he called *za-labia*. His neighbor on the pike was an ice cream vendor. One hot day the ice cream man ran out of dishes and Hamwi came to his rescue by rolling his waffles into cone shapes. This ice cream in a cone was dubbed the World's Fair Cornucopia and became a sensation.

(ii) Enough containers to provide a total quantity of declared weight of at least 24 pounds if they bear a weight declaration of more than 16 ounces.

(2)Determine net weight of each frozen pie.

(3)Temper the pie until the top crust can be removed.

(4)Remove the filling and cherries from the pie and transfer to the surface of a previously weighed 12-inch diameter U.S. No. 8 sieve (0.094-inch openings) stacked on a U.S. No. 20 sieve (0.033-inch openings).

(5)Distribute evenly over the surface and wash with a gentle spray of water at 70 degrees–75 degrees F. to free the cherries and cherry fragments from the adhering material.

(6)Remove the U.S. No. 8 sieve and examine the U.S. No. 20 sieve and transfer all cherry fragments to the U.S. No. 8 sieve.

(7)Drain the cherry contents on the No. 8 sieve for 2 minutes in an inclined position (15 degree–30 degree slope). Weigh the U.S. No. 8 sieve and the washed and drained cherries to the nearest 0.01 ounce.

(8)The weight of the washed and drained cherries is the weight of the sieve and the cherry material less the weight of the sieve. Calculate the percent of the cherry content of each pie with the following formula, and then calculate the average percent of the entire random sample:

Percent of the cherry content of the pie =

Weight of washed and drained cherries X 100

Net weight of pie

(c)If the quality of the frozen cherry pie falls below the standard of quality prescribed by paragraph (a) of this section, the label shall bear the general statement of substandard quality specified in 10.7 (a) of this chapter in the manner and form specified therein; but in lieu of the words prescribed for the second line inside the rectangle, the label may bear the alternative statement "Below standard in quality—," the blank being filled in with the following words, as applicable: "too few cherries," or "blemished cherries." Such alternative statement shall immediately and conspicuously precede or follow, without intervening written, printed, or graphic matter, the name of the food as prescribed by 28.1.

What does all that government-ese mean? That a cherry pie has to contain cherries. That 25 percent of the weight of the pie must be the cherries. That they have to be top-grade cherries, well, mostly top-grade cherries: Fifteen percent can be seconds. That Banquet can't put a picture of the prize-winning cherry pie from the state fair on the box if the pie inside doesn't come close to looking like that. That Banquet can put just about anything except artificial sweeteners inside the pie as long as the additives are federally approved. And if it is a small pie, Banquet can call it a cherry tart, but it doesn't have to.

Candy, Cookies, Paper Products

▪▪▪▪▪▪▪▪▪▪▪▪▪▪▪▪▪▪▪▪▪▪▪▪▪▪▪▪▪▪▪▪

The candy industry has a goal: "25 by '95."

By 1995 the country's confectioners want the per capita consumption of candy by Americans to reach 25 pounds. It's already 20 pounds, up from 16.1 pounds per person in 1980.

What does 25 pounds figure out to? Well, your basic Snickers Bar weighs in at 2.07 ounces. To eat 25 pounds of these a year, you'd need to eat 193, better than one every other day. Hey, I'm carrying my share of the load.

The candy companies have a plan to reach this goal, too: Add a new national holiday, Candy Carnival Day. (This is true.) It would be sometime in May, which is already National Candy Month, and it would help sell candy in that barren candy stretch between Easter and Halloween.

Sorry, dentists. It's probably not going to happen, "25 by '95," that is. Even by the most optimistic of estimates, the one formulated by the National Confectioners Association, we won't reach that level of chocolate-covered-gluttony until 1998. U.S. Commerce Department projections place it even later, 1999.

6-COUNT PACKAGE HERSHEY'S MILK CHOCOLATE BARS—$4.95

Would you buy a candy bar if you knew that the middle name of its inventor was Snaveley? If not, you'd better skip this section.

Candy consumption has come a long way in the hundred years since Milton Snaveley Hershey invented the candy bar, way back in 1894. Oh, there were chocolate bars before Hershey's. Most people credit the British firm of Fry and Sons with making the first chocolate bar, sometime in the 1840s. But it was left to Hershey to invent the *candy bar*.

Hershey got the idea when he saw a German chocolate-making machine at the 1893 Chicago World's Columbian Exposition. He was intrigued. He already made caramels at his Lancaster, Pennsylvania, factory. He ordered one of the German machines and began experimenting. The next year he loosed on the market a couple of candy bars that are still popular today: the Hershey's Milk Chocolate Bar and the Hershey's Milk Chocolate with Almonds Bar.

The success of these first two candy bars brought competitors storming into the market. The first peanut bar, a combination of peanuts and chocolate, was introduced in 1905 by Perley Gerrish of Cambridge, Massachusetts. One of the first candy bars to use multiple ingredients was the still-popular Goo Goo Cluster, invented in Nashville in 1912. It combined caramel, coconut, marshmallow, milk chocolate, and peanuts.

There were hundreds of variations on the nut roll—peanuts, caramel, and fudge coated with chocolate—from the Baby Ruth Bar to Oh! Henry. Contrary to popular belief the Baby Ruth, which was introduced by Curtiss Candy Company of Chicago in 1920, was not named for the baseball player but for former president Grover Cleveland's daughter, who had been the nation's sweetheart in the 1890s. But after Babe Ruth became a superstar, he lent his name to a couple of other candy bars: the Bambino and the Big Champ.

The next breakthrough in candy bar history came in the early 1920s and, as with many great inventions, it was an accident. Technicians at the Pendergast Candy Company in Minneapolis put too much egg white into the nougat recipe for a new chewy candy bar they were developing, the Emma bar. What was supposed to be chewy came out fluffy, and they promptly renamed the bar Fat Emma. Frank Mars of Chicago copied the fluffy nougat center for his new candy bar, the Milky Way, which arrived in stores in 1923.

That same year H. B. Reese of Hershey, Pennsylvania, came out with a variation on the peanut roll, coating peanut butter instead of peanuts, and Reese's Peanut Butter Cups were born.

Snickers, which was a Milky Way with peanuts, was introduced in 1930 by Mars. Two years later Mars came back with another variation on the Milky Way. This time he put three different flavors in the nougat center, wrapped three small bars in one package, and called it 3 Musketeers.

The candy bar took another giant leap forward in the thirties when Frank Martoccio of Minneapolis developed a synthetic coating for chocolate that would stay hard in hot weather. He went on to produce Zero and Milk Shake, two candy bars that are still around, if not necessarily thriving.

16 OZ. BAG M&M'S PLAIN—$2.39

The first M stands for Mars, Frank Mars, founder of Mars Candy; the second for Murrie, Bob Murrie, the president of Mars Candy in 1941.

Yes, 1941, the year that gave us Pearl Harbor, also gave us M&M's, the multicolored candy that melts in your mouth, not in your hand.

There are six colors in an M&M's pack: brown, yellow, red, green, orange, and tan. Red M&M's were dropped in 1976 during the red dye number 2 scare. They didn't contain any red dye number 2, but Mars was afraid that consumers might think they did. Red M&M's returned, with much fanfare, in 1986.

You could stay up all night counting how many of each color are in a bag. Or you could read on.

Here's the current breakdown by color. It's based on extensive research by Mars and changes periodically.

	Plain	Peanut
Brown	30	30
Yellow	20	20
Red	20	20
Green	10	20
Orange	10	10
Tan	10	0

You tell me how you conduct research on something like that. Do you put rats in a maze and see which M&M they go to? Or do you interview people: "Uh, let's say there were a hundred M&M's in a bag, how many do you think should be red? How about in a bag of M&M's Peanut?"

There's no difference in flavor among the colors, but I guess you already knew that.

II OZ. PACKAGE DUNCAN HINES SOFT & CHEWY CHOCOLATE CHIP COOKIES—$1.78

This is the cookie that launched the Great Cookie War of 1984. It was probably the fiercest, most expensive, and ultimately most disappointing war ever waged for the grocery shelf.

The cookie war began innocently enough in Kansas City in 1982. That's when Frito-Lay, the potato chip king, quietly tested a new cookie from Grandma's Cookies, an Oregon brand it had purchased in 1980. This was startling enough because Frito-Lay was known for salty snacks. It had never marketed cookies before. The cookie shelves of the grocery store—at the time a $2.1

billion market—belonged to Nabisco (Chips Ahoy!, Oreo), Kee-bler, and Sunshine.

Then a few months later, in January 1983, Procter & Gamble, the soap kings, a company that in its then 145-year history had never sold a packaged cookie, stormed into Kansas City with a new Duncan Hines brand, Duncan Hines Soft & Chewy Cookies.

This was no overnight assault. P&G food scientists had labored for seven years trying to come up with a way to make cookies that were soft on the inside and crisp on the outside and would stay that way through shipping and onto the shelf. Duncan Hines Soft & Chewy Cookies were more than just cookies; they were a concept. Cookies that were just like homemade.

Despite the longtime success of Chips Ahoy! and Oreo and the various Keebler soft cookies, Procter & Gamble thought there was another market out there, one for fresh-tasting cookies. It pointed to the success of in-store bakery cookies as evidence.

Procter & Gamble's goal was to produce a soft-centered cookie that would taste like the ones Grandma used to make. But Grandma couldn't duplicate this cookie: It required three different ovens and such strange ingredients as polyglycerol esters and diacetyl tartaric esters of mono- and diglycerides. Grandma would have had to have a food science degree and also infringe on a couple of patents.

The reason store-bought packaged cookies didn't taste like Grandma's homemade cookies was the texture. While cooking, the sugar in cookies caramelizes, giving them that special chewy texture. But after the cookie cools, the sugar reverts to its crystal state and the cookie has a different texture, a harder texture.

The packaged cookie market had been in decline for ten years when P&G made its move. In 1972 Americans consumed 2.3 billion pounds of packaged cookies. By 1982 that had dropped to 2 billion pounds.

Part of the reason for the drop was a changing population. The number of kids aged five to thirteen—prime cookie-eating age—had declined. Meanwhile, the nutrition concerns of mothers had increased. In addition, part of the market was being nibbled away by fresh-baked cookie stores such as The Great American Chocolate Chip Cookie Company, David's Cookies, and Mrs. Field's Cookies.

Procter & Gamble also knew that the big-three cookie makers were vulnerable to an all-out assault. They had become complacent over the years, spending only $15 million combined on cookie advertising in 1982. Frito-Lay chairman D. Wayne Calloway would later tell stock analysts his company spent that much on Doritos alone.

Procter & Gamble first applied for a patent on its new cookie dough in March 1981. It was granted June 19, 1984: Patent No. 4,455,333.

The mistake P&G made with this new cookie was in taking the safe path to market, by testing the cookie in selected stores before rolling it out nationally.

To test or not to test is always the question. Every giant food company must confront that dilemma when it comes time to introduce a new product. If the company tests the product, it finds out if consumers will buy it; it discovers the "bugs" in the product, the little things that can be worked out in a test market rather than facing embarrassment on a national scale; and it reduces the risk of rolling out a doomed product. But it also runs the risk of losing a competitive edge.

While Procter & Gamble was test marketing its soft cookie in Kansas City, its competitors were scrambling to come up with a similar product. And in the case of the Duncan Hines Soft & Chewy line, P&G lost the advantage it would have gained by jumping headfirst into the new product pool.

In the last few years companies have shortened their test market times. And many of the second-place companies in various segments—the ones, like Avis, that have to try harder—have skipped test marketing new products altogether.

In the fifties the big food companies tested virtually every new product. The company would settle on a test city—Columbus, Ohio, was a favorite in the fifties—then place the product in supermarkets around town. There might also be a splurge of TV and newspaper advertising to introduce and support the new line.

Then after six months or so, company executives would look at the numbers: How did the product sell? How did it sell compared to comparable items? Were they able to keep costs in line with charges?

If the product passed the test, it would be put into national distribution, with a national marketing campaign.

The main reason testing is losing favor is competition.

A company may spend hundreds of thousands of dollars developing a new product, then more money testing its appeal in retail stores, only to have a competitor "steal" the product.

So back in late 1982 Nabisco's officers blithely forecast increased earnings in 1983 and 1984. Then Duncan Hines Soft & Chewy Cookies hit the shelves in Kansas City and all hell broke loose. Nabisco moved fast, cutting its earnings estimate by $50 million, and spending roughly that amount to market its own brand of soft cookies—Almost Home. Keebler moved almost as quickly, launching its own Soft Batch Cookies.

And before Duncan Hines cookies were even in the national market, the battle was lost. Nabisco and Keebler had virtually snuffed out Duncan Hines cookies.

A loss on the grocery shelves only stiffened P&G's resolve. It had forecast this soft cookie market, worked diligently to develop the cookie, and then saw its competitors copy its product. In June 1984 Procter & Gamble went to court, suing Nabisco, Keebler, and Frito-Lay for violating its patent. It also accused its three rivals of "unfair competitive practices." That's legalese for industrial espionage.

P&G claimed Nabisco figured out its process by sending a spy to a contract manufacturer where P&G's "secret cookie-making technology was being used." It accused Frito-Lay of directing an employee to pose as a "supervisor of a potential customer" and attend a confidential sales presentation.

Worst of all, it claimed Keebler, the home of those cute little elves, had "rented an airplane and took aerial photographs of Procter & Gamble's Jackson, Tennessee, cookie manufacturing facility while its construction was in progress."

Frito-Lay admitted it had a representative photograph of the

new Duncan Hines bakery exterior but blamed the spying inci-
dent on the man's college-age son, who entered the factory on
his own and asked for unbaked dough. Frito-Lay assured P&G
it had destroyed both the dough and the photos without so much
as a long look. Right.

This is where the war really got interesting.

Nabisco hired an army of researchers to scour the universe of
published cookbooks, looking for a two-dough cookie, one that
might have the same characteristics as the Duncan Hines cookie.
If they could come up with a previously published recipe, they
could invalidate P&G's patent and the lawsuit would be null
and void.

They finally found such a recipe in a 1968 Canadian cookbook titled *Food that Really Schmecks* by Edna Staebler. (Honest, that was the name of it.) It was an Old Order Mennonite
recipe for Rigglevake Cookies—Railroad Cookies.

Staebler soon became the discreet object of desire of both
companies, P&G and Nabisco. And Old Order Mennonite women
became even more sought after. Both companies were writing
out checks to little old women for a day's work in the kitchen.
Lawyers were shuttled in and out, researching the origin of the
recipe.

There was only one small problem. Their religion forbade the
Mennonite women from testifying in court. No amount of ca-
joling—or wooing—could change that. And in the end the lawyers
abandoned the Canadian connection.

The patent infringement case was finally settled in September
1989, five years after the Great Cookie War had begun. Nabisco
Brands, Inc., Keebler Company, and Frito-Lay, Inc., paid a total
of $125 million, the largest settlement ever reported in a patent
case.

But the postscript had already been written before the case
was settled: A December 1985 *New York Times* headline said it
all: "Chewy Cookie Market Falters."

Soft cookies never took off. Cookie lovers just never believed
that store-bought cookies could be as good as home baked.

In June 1987 P&G cut back production and promotion of Duncan Hines Soft & Chewy Cookies. Industry insiders estimated the cookies had cost the company $100 million.

P&G won the battle; everyone lost the war.

12 OZ. BOX NILLA WAFERS—$2.39

You know the name. I know the name. It's the cookie we grew up with: Nilla Vanilla Wafers.

Except that's not the name anymore.

Just as Sugar Crisp has become Golden Crisp and Sugar Corn Pops has become Corn Pops, so too has Nilla Vanilla Wafers changed its name. But not for fear of some backlash against the use of vanilla.

Kellogg's and Post erased "Sugar" from their cereal names but not from their ingredient lists. Nilla Vanilla Wafers, on the other hand, is now simply Nilla Wafers because the cookies no longer contain vanilla. And to qualify to have "Vanilla" in the name, it would need vanilla in the wafers. But vanilla has become too scarce. There isn't enough produced in the world to flavor all the vanilla ice cream, much less the vanilla wafers. And because of the scarcity, vanilla has become expensive. So in the eighties Nabisco changed the vanilla ingredient to vanillin, an artificial flavor, and quietly dropped "Vanilla" from the product name.

16 OZ. BOX HONEY MAID
HONEY GRAHAMS—$2.08

The health food craze didn't begin with joggers and granola groupies in the sixties or with cholesterol counters and fat-free fanatics in the eighties.

It began with religious fanatics in the thirties, the 1830s!

And one of the leading food faddists was a onetime agent for the Pennsylvania State Society for the Suppression of the Use of Ardent Spirits, an ordained Presbyterian minister named

Sylvester Graham. He was an early advocate of natural food, at a time when most people thought all foods were equal, the only difference being in the quantity consumed. Graham preached the gospel of God's food, unadulterated by mammon: a gospel that has been preached by food faddists throughout the ages, from the Kellogg brothers of Battle Creek, the first food faddist millionaires, to Jerome Rodale, publisher of *Prevention* magazine, and food writer Adelle Davis. Davis denounced refined sugar, pasteurized or homogenized milk, white bread, and food additives, and claimed the Nazis overran France so easily because the Germans dined on black bread and beer instead of the Frenchman's white bread and wine. Davis died at age seventy of bone cancer.

But the first, and possibly the most influential of all, was Graham. By 1838 the country was overrun by "Graham boarding houses," rooming houses that served meals in accordance with Graham's teachings. He opposed any alteration of foods from the way the Creator had made them, which meant lots of raw fruits and vegetables and very little meat. Meat, he believed, was the devil's doing, a food that aroused both the temper and the loin.

Graham was particularly opposed to the flour fashionable at the time, white flour. He believed it to be against the teachings of the Bible to remove the bran from the wheat before milling and thus lent his name to Graham flour, a brown meal made by

---

FORKLORE #25

HEALTH FOOD

The man who invented Twinkies, James Dewar, lived to be eighty-eight years old. He attributed his long life to the fact that he ate two Twinkies a day, every day, from the time he invented them in 1930.

I wonder how long he would have lived if he hadn't eaten them?

---

crushing the bran with the wheat. And from Graham flour came the Graham cracker, a slightly sweet, honey-colored, flat cracker.

So the humble graham cracker—the purest of children's food—had its origins rooted in sex and sin, having been created as an anti-lust food. And today we use graham crackers in the crusts of such sinful creations as cheesecake and Key lime pie.

20 OZ. PACKAGE OREO COOKIES—$2.39

This is America's favorite cookie.

But when the National Biscuit Company introduced the Oreo in 1912 in Hoboken, New Jersey, it certainly didn't foresee that it would catch on like it has. In fact, the company had higher hopes for two other cookies that made their debut at the same time. In an April 2, 1912, internal memo to its managers the National Biscuit Company (*biscuit* is the British word for "cookie") said, "We are preparing to offer to the trade three entirely new varieties of the highest class biscuit packed in a new style. The varieties are as follows:

"Mother Goose biscuit—a rich high class biscuit bearing impressions of the Mother Goose legends.

"Veronese biscuit—a delicious, hard sweet biscuit of beautiful design and high quality.

"Oreo biscuit—two beautifully embossed, chocolate-flavored wafers with a rich cream filling."

It was as if this new Oreo were an afterthought—not "delicious" or "rich high class" but only "embossed."

So when was the last time you had a Mother Goose or a Veronese? Those two died out quickly, but the Oreo has endured. Nabisco estimates it has manufactured 200 billion Oreos since 1912. That's 2.5 billion a year for eighty years.

Or to use a measurement we all understand: That's 8 trillion calories. If one person had eaten every one of those Oreos—and I know a couple of people who'd like to try—he or she would have gained 400 million pounds, the equivalent in weight of the population of Philadelphia.

No one at Nabisco is sure where the name Oreo came from. There are two theories: Some say it came from the Greek word for mountain. Adolphus Green, who was head of National Biscuit in 1912, was known to be a classics scholar. More likely it came from the French word for gold. The first label had scrollwork in gold on a pale green background with the product name in gold.

Over the years the Oreo's size has fluctuated. The current cookie, which is an inch and three-quarters in diameter, is about in the middle of the variations.

In 1991 Nabisco introduced the Mini Oreo, which is one inch across. The Mini Oreo was a latecomer to the mini-snack market that began in 1987 with Ritz Bits.

But there was good reason for the delay. Nabisco had to make sure Oreo lovers could still pull the top off to eat the creme first.

What's Oreo's secret? No secret really. Just quality chocolate in the cookie (at one time Nabisco bought their chocolate from the same people who supplied Godiva Chocolates) and sugar and shortening in the filling.

10-COUNT BOX MOON PIES—$1.59

Where I come from, used to be every crossroads had a store. And every store had a gang of regulars, grown men who called themselves boys, who would sit out front and discuss the world situation. Pickup trucks would whistle by and the boys would throw up their hands in a wave. Every so often one of the boys would hoist himself up out of his hind-catcher crouch and lumber inside to buy himself something to drink and something to munch on. Often as not, something to drink was an RC. And something to eat was a Moon Pie.

RC and a Moon Pie. The common man's banquet in the South in the thirties, forties, and fifties.

The RC and Moon Pie became a celebrated meal during the heyday of King Cotton. RC, or Royal Crown Cola, as it is more correctly known, was the first five-cent soft drink in a ten-ounce bottle. And the Moon Pie was the largest five-cent snack cake on the market.

So for ten cents, field workers and farm boys could have a feast during their lunch break, a dime's worth of calories that would last an afternoon. And fill them up to boot.

RC and a Moon Pie became an inseparable pair, in the country stores and in the language. They reigned in the South for almost half a century. But then they quit building crossroads and started building intersections. They erected interstate highways that didn't just pass near communities, they passed them by. And things started changing in the soft drink and baking industries. RC went national and supermarkets replaced corner groceries.

Chero-Cola Company, the manufacturer of Royal Crown Cola, sold out to the Nehi Company in 1928 and Royal Crown expanded into the national market, taking aim at the two giants, Coke and Pepsi, in a battle still being waged, still unsuccessfully. In 1990 RC ranked eleventh among cola drinks, with a mere 1.6 percent of the market. And that was down from 2.8 percent in 1980.

While RC was expanding, the Moon Pie was content to be a regional pastry. The Chattanooga Bakery just kept turning out Moon Pies for the South and the South alone.

That's changed in the last ten years. While RC has been in eclipse, the Moon Pie has been rising. Available for years in only about fourteen states, it is now in stores in forty of the fifty. And longtime Moon Pie lovers in the other ten states send regular missives, begging the company to sell Moon Pies in their area.

Two billion Moon Pies have been eaten since their creation in 1917. That's crumbs next to Oreos. But that's a lot for a little bakery.

The progenitor of the Moon Pie was a snack cake called the Lookout Marshmallow. It was described in a Chattanooga Bakery sales brochure as a "delicious sandwich, marshmallow between vanilla wafers."

The Moon Pie was created when the bakery replaced the vanilla wafers with a cookie closer in taste to a graham cracker.

The early cakes were called Lookout Moon Pies and sold for five cents.

There are two versions of how the Moon Pie got its heavenly name. In one, the Chattanooga Bakery's sales manager dubbed it the Moon Pie because it looked like the moon. The other has a Knoxville field manager named Earl Mitchell reporting on a sales trip through Appalachia: "They don't want anything that we got. They want something big and round, filled with marsh-mallow and covered with lots of chocolate, and it needs to be as big as the moon."

At the time of its creation the Moon Pie was but one of 200 different cakes manufactured by the bakery. But it soon became the star confection. During the next four decades, the list of products dwindled and for the last thirty years or so the Moon Pie has been the only cake made at the bakery's plant on King Street in downtown Chattanooga.

The bakery turns out 300,000 Moon Pies every day.

The Moon Pie is no longer a nickel, but it's cheaper—and big-ger and more filling—than a candy bar. How do they keep the price of this legendary cookie down?

They don't advertise.

12 OZ. BOX RITZ CRACKERS—$2.38

National Biscuit Company created the Ritz Cracker in 1934 by increasing the amount of shortening in a regular cracker, re-moving all the yeast, and then adding a thin coat of coconut oil and a sprinkling of salt. It was named Ritz to give it an air of class.

6.75 OZ. PACKAGE PEPPERIDGE FARM STRAW-BERRY FRUIT COOKIE—$1.65

I don't usually read food labels in the grocery store: thirty-four percent of shoppers don't. But while Judy tries to make her cookie selection, I'm killing time reading all the drivel on this fancy cookie. Perhaps the most unusual notation is this: "Reg. Penna. Dept. Agr." I've seen this before. But why does "Pepperidge Farm, Inc. Gen. Off., Norwalk, CT" have to register with the Pennsylvania Department of Agriculture? Why not Connecticut or, better still, Louisiana, the Sugar State?

In order to sell baked goods in the state of Pennsylvania, a company must be "registered with the Pennsylvania Department of Agriculture." This is due to the Pennsylvania Bakery Law of 1933, which required all baked goods—and the definition of baked goods is so broad as to include spaghetti, macaroni, even potato chips—sold in the state to pass rigorous standards for cleanliness and fair weight. Most manufacturers incorporate the "registered" disclaimer into their standard box design rather than having a separate box for products sold in Pennyslvania.

Pennsylvania has reciprocal agreements with inspectors in other states so that Pennsylvania food inspectors don't have to comb the countryside, on the lookout for short-weighted Cheez-It packages and Eagle Brand Potato Chips that have settled a little more than necessary during shipping.

AISLE 9

Soft Drinks

▲▼▲▼▲▼▲▼▲▼▲▼▲▼▲▼▲▼▲▼▲▼▲▼▲▼▲▼▲▼

Judy carefully loads three 2-liter soft drink bottles into the bottom rack on our cart. Three different drinks because the adults want diet, the kids want regular. She expresses her dismay: "I wish they still carried soft drinks in glass bottles. I could get two six-packs and mix different drinks in them. Now I have to buy either a two-liter bottle or a six-pack of cans."

For a half century soft drinks were sold in grocery stores in glass bottles. Then in the fifties soft drink companies began experimenting with cans. The initial resistance—"It tastes like the can"—was overcome and by 1980 cans accounted for the majority of soft drink sales. But even that passed as soft drink companies introduced the two-liter family-size plastic bottle.

No food has invented and reinvented its packaging more in my lifetime than the soft drink.

The Coca-Cola case at Joyner's Grocery was a gathering spot for day laborers during their lunch break. Grown men would reach down into the watery ice, pluck out a 6 1/2-ounce bottle of Coke, pop the lid off on the opener mounted on the side, and take a swig straight from the bottle.

In 1953 the returnable glass bottle—two cents deposit per bottle—reigned supreme in the soft drink case. As it had since the turn of the century when soda pop went from healthful beverage to refreshing beverage.

But in the forty years since I first stepped into Joyner's, I have seen Coke in 12-ounce tin cans, Coke in 12-ounce aluminum cans, Coke in 16-ounce returnable bottles, Coke in 32-ounce "family-size" glass bottles, Coke in 64-ounce "family-size" glass bottles, Coke in 16-ounce throwaway bottles and, every Christmas, Coke in old-style 6 1/2-ounce throwaway bottles. And now there's Coke in the 2-liter plastic throwaway bottle.

Soda pop began as a bottled drink because all commercial drinks in the nineteenth century were bottled drinks: milk, beer, mineral water.

The tiny 6 1/2-ounce bottle was the standard. But the soft drink market was hit hard by the Depression. Pepsi fought back in 1934 by introducing the 12-ounce bottle for the same price: "Twice as much for a nickel," the ads trumpeted. Coke did not follow suit. In fact, it wasn't until 1955 that Coke broke with its fifty-six-year tradition of one flavor, one size, and offered Coke in the 12-ounce bottle.

Continental Can tried to interest soft drink manufacturers in switching to cans as early as 1936. Continental even test marketed Clicquot Club Soda in cans that year, but there were problems: The cans leaked and customers complained that the soda had a metallic taste. But the biggest problem was with display. The cans weren't flat on top—they had a bit of a bubble—and they were hard for grocers to stack.

Continental and Pepsi tried to introduce canned soda pop again in 1948, but customers weren't willing to pay the premium price for the convenience of a can. The next year Pepsi dropped the experiment.

Canned soft drinks made their official debut in 1953 with the drink Super Cola. But it was still a small market. All the kinks hadn't been worked out. The can was difficult to open—there was much more foaming than with beer—and there were still complaints about a metallic taste. Grocers didn't like them because the inferior linings were giving the drinks a short shelf life.

Two events pushed canned soft drinks into the mainstream. The first was the introduction of the pull-tab can by Alcoa in 1960. No longer would the consumer have to fish under the car seat for that missing church key in order to open a soft drink. Now he or she just pulled a metal ring and—presto—there was a drinking spout. And a metal ring to get rid of.

The second was Reynolds Aluminum's development of the all-aluminum can in 1963. It was lighter in weight and there was no metallic aftertaste.

The canned soda pop boom began.

Nineteen sixty-one marked the first year that all the major brands—Coke, Pepsi, 7Up, Canada Dry, RC, Dad's, Dr Pepper, Hires, and Bubble Up—were available in cans. The first soft drink to use the all-aluminum can was Slenderella Diet Cola in 1963. RC switched to aluminum cans in 1964, Pepsi followed in 1967, and Coke followed suit the next year.

The nonremovable opener—the pop top—was introduced in 1974 in response to safety and environmental complaints about the ring tabs. Ring tabs were turning up everywhere, from national parks to fish bellies.

But even as soft drink lovers were adapting to cans, the soft drink companies were working on new packages. One of these

---

FORKLORE #26

HOLY SH——!

During World War II, when caffeine was scarce, Coke executives discussed using synthetic caffeine from bat guano (that's bat do-do) but rejected the idea because they knew if word ever got out that Coke had bat excrement in it, they might as well turn over the plant keys to Pepsi.

---

was the larger family-size glass bottles. The first ones held a quart, about as much as two large bottles. Next came the half-gallon size. A new coating called plasti-shield, developed in 1971, made the large glass bottles possible. It covered most of the glass and made the bottle shatter resistant, if not shatterproof.

The sixty-four-ounce glass bottle was a success in the supermarket but not in the accounting department. *Beverage World Gazette* reported in 1977 that the bottle was costing soft drink companies one-and-a-half times more than the ingredients in the bottle.

But the soft drink companies were planning ahead. They had been working on a two-liter plastic bottle and in 1978, with much fanfare, Coca-Cola and Pepsi introduced this new acrylonitrile plastic bottle. And in no time the Food and Drug Administration banned the bottle because acrylonitrile was a carcinogen.

On the heels of that fiasco the soft drink companies brought out a new polyethylene terephthalate (PET) bottle. It too had its problems *Beverage World Gazette* noted that when stored above 100° F. the drinks had a pronounced tendency to blow their lids.

But there was no arguing with the idea. In 1977 there were zero 2-liter plastic bottle sales. In 1978 that jumped to 600 million bottles sold. In 1979 it leaped to 1.6 billion. And by 1980, sales of 2-liter bottles had topped 2 billion. It was an overnight success.

The liter unit of measurement was adopted because America was changing to the metric system. When America decided it wasn't going to change to the metric system after all, soft drink companies stayed with the liter because the term "two-gallon bottle" reminded shoppers of gasoline and no one wanted to drink that.

2-LITER BOTTLE COCA-COLA CLASSIC—$1.49

This is it. The mother of all soft drinks.

Coca-Cola's origins can be traced to the 1885 experiments of an Atlanta chemical company owner named John Styth Pemberton. Pemberton, who had already created Globe of Flower Cough Syrup, Indian Queen Hair Dye, Triplex Liver Pills, and the blood medicine Extract of Styllinger, was trying to copy a popular health elixir of the time, Vin Mariani, a wine spiked with coca. Pemberton's concoction, which he dubbed French Wine of Coca, was designed to be a stimulant, a health drink like the kind favored by such noteworthy folk as Ulysses S. Grant and Jules Verne. Pemberton advertised it as an "ideal nerve tonic and stimulant."

But French Wine of Coca was a flop. So Pemberton went back to the laboratory drawing board. He added kola nut extract because it contained the stimulant caffeine and substituted flavored sugar water for the wine. The result was a thick syrup with a sweet taste and a bit of a kick. It could be consumed straight or mixed in water. Pemberton's bookkeeper, Frank M. Robinson, gave the drink its name: Coca-Cola, changing the *k* in *kola* to a *c* to enhance the script logo he had in mind.

Pemberton took his new syrup drink to druggist Willis Venable at Jacob's Drug Store. It was Venable who added carbonated water, either on his own or at the suggestion of a customer. No one is sure which. During the first year Venable sold an average of thirteen glasses of Coca-Cola a day.

FORKLORE #27

IT'S THE REAL PAUSE THAT CAN'T BEAT THE FEELING

Coca-Cola has had more memorable advertising catchphrases than any other company. But let's be honest: It didn't start out that way. Here is a guide to the changing Coke slogans:

    1886 — Drink Coca Cola.
    1891 — A delightful summer or winter drink.
    1900 — Good to the last drop. (Coke dropped this slogan;
           Maxwell House Coffee picked it up.)

1904 — Delicious and refreshing.

1905 — Coca-Cola revives and sustains.

1906 — The great national temperance beverage.

1915 — Pause in the mad rush and seek a soda fountain.

1917 — Three million a day.

1922 — Thirst knows no season.

1925 — It had to be good to get where it is.

1927 — Around the corner from everywhere.

1929 — The pause that refreshes.

1932 — Ice-cold sunshine.

1936 — The refreshing thing to do.

1938 — The best friend thirst ever had.

1939 — Coca-Cola goes along.

1939 — Wherever you are, whatever you do, wherever you
       may be, when you think of refreshment think of
       ice-cold Coca-Cola.

1942 — The only thing like Coca-Cola is Coca-Cola itself.
       It's the real thing.

1948 — Where there's Coke there's hospitality.

1949 — Coca-Cola . . . along the highway to anywhere.

1952 — What you want is a Coke.

1956 — Coca-Cola . . . makes good things taste better.

1957 — Sign of good taste.

1958 — The cold, crisp taste of Coke.

1959 — Be really refreshed.

1963 — Things go better with Coke.

1970 — It's the real thing.

1971 — I'd like to buy the world a Coke.

1975 — Look up, America.

1976 — Coke adds life.

1979 — Have a Coke and a smile.

1982 — Coke is it!

1985 — We've got a taste for you.

1985 — America's real choice.

1986 — Catch the wave.

1987 — When Coca-Cola is part of your life, you can't beat
       the feeling.

1988 — Can't beat the feeling.

In July 1887 Pemberton and Robinson sold the formula for this new health tonic to Venable and his partner, George S. Lowndes. They owned the rights—and the all-important formula— for only five months before selling out to Woolfolk Walker and Mrs. M. C. Dozier. Obviously none of these folks knew what they had because Walker and Dozier turned around and sold the name and formula in 1888 to Asa G. Candler.

It would be Candler who would turn Coca-Cola into a household name. He did a little tinkering with the sacred formula, by all accounts improving the taste of the drink. William Poundstone, in the book *Big Secrets*, speculates that Candler added phosphoric acid, among other changes. He also turned the formula into a closely guarded secret. Only Candler and his new partner, Frank Robinson—the same—were allowed in the lab when the formula was being concocted. Ingredient labels were removed as bottles arrived. Only Candler was allowed to open the mail; only Candler was allowed to pay the bills. The name Coca-Cola was trademarked on January 31, 1893. In 1945 the company trademarked the name Coke. But because it didn't trademark the second part of its name, "Cola," the term has become synonymous with an entire category of dark brown soft drinks.

Shortly thereafter Candler took out an ad in the *Atlanta Journal*:

"Coca-Cola.

"Delicious. Refreshing. Exhilarating. Invigorating.

"The new and popular soda fountain drink containing the tonic properties of the wonderful coca plant and the famous cola nuts on draught at the popular soda fountains at 5 cents per glass."

Also in the ad was a pitch for Candler's other product, Delectalave, the Great Tooth Wash.

By 1894, sales of Coca-Cola were too large to handle production of the syrup in one central location. That's when Candler devised his numbered ingredients system. Instead of giving the formula to each branch plant, he shipped the ingredients to them in numbered bottles. They were to mix up the formula, using bottles 1 to 9, according to a recipe provided from Atlanta headquarters.

But the Coca-Cola secret formula was not as secret as Candler would have liked. An Atlanta woman named Diva Brown claimed that John Pemberton sold her the Coke formula shortly before he died in 1888, and over the next couple of decades she sold passable imitations under the names Better Cola, Celery-Cola, Lime Cola, My-Coca, Vera-Coca, Vera Cola, and Yum-Yum. She died in 1914 but not before selling her secrets to two other manufacturers, businessman John Fletcher and the Koke Company. Both were eventually put out of business by lawsuits from Coca-Cola.

In 1914, with imitation cola drinks running rampant, Candler decided Coke needed its own distinctive bottle, one that Coke lovers could recognize in the dark. To accomplish the task he went to Don Samuelson at Root Glass in Terre Haute, Indiana. Samuelson looked in the *Encyclopaedia Britannica* for a picture of a kola nut and created the now-famous green bottle to match the egg-shaped nut.

Coca-Cola came under scrutiny from government watchdogs in 1909. Government agents seized a shipment of Coke syrup bound from Atlanta to Chattanooga and charged the company under the recently minted Pure Food and Drug Act of 1906. The government believed that Coca-Cola contained the recently banned substance cocaine. There was also concern about the name. If it was called Coca-Cola, it should have coca and kola in it. If it didn't, then it was misbranding, also a violation of the Pure Food and Drug Act.

During trial in the case of *United States v. Forty Barrels and Twenty Kegs of Coca-Cola*, the government presented several chemical analyses of Coca-Cola, the closest official reading we have to date of what is in a Coke. By one analysis Coca-Cola was 48.86 percent sugar. Water was pegged at anywhere from 34 to 41 percent. Also present in detectable amounts were caffeine, phosphoric acid, caramel, glycerin, lime juice, and oil of cassia (Chinese cinnamon).

No coca, no kola.

That's when Coca-Cola brought in one Dr. L. Schaeffer, president of Schaeffer Alkaloid Works of Maywood, New Jersey. He testified that he manufactured what Coca-Cola called Merchandise No. 5 and it was made from coca—with the cocaine removed—and kola nut dissolved in a solution of alcohol and California wine.

No cocaine, but yes, coca and yes, kola. And Coca-Cola was absolved of any criminal wrongdoing.

But it still leaves unanswered the question, What is in Coke? According to William Poundstone, Coke is 99.5 percent sugar water. What else is in there? Caramel made from burned cane sugar or burned corn sugar gives Coke that characteristic brown color. Caffeine gives it its jolt. Phosphoric acid is responsible for the tang. Lemon oil, orange oil, lime oil, cassia oil, and nutmeg oil—all dissolved in a 95 percent alcohol solution—is where the flavor comes from. Glycerin is a preservative. Vanilla extract gives Coke a touch of cream soda taste. And coca leaf and kola nut—in a 20 percent alcohol solution—are there only so you can call it Coca-Cola. Poundstone says the drink tastes the same with or without the coca and kola ingredients.

There's no need to detail the fiasco of New Coke in this book. Suffice it to say my Winn-Dixie carries five rows of Coca-Cola Classic and one row of New Coke. And with good reason. Despite the largest rollout and most publicity of any reformulated food product ever, it fizzled. It now has a 1.5 percent share of the total soft drink market, trailing even RC.

2-LITER BOTTLE TAB—$1.49

Today I'm also picking up a bottle of Tab. If I can find it.

My cousin is coming to town and she is a Tab lover.

Of all the well-known soft drink brands, Tab is the hardest to find. Coca-Cola claims it is in 92 percent of all supermarkets and that Louisville is one of the big three for Tab (Memphis,

Tennessee, and Lexington, Kentucky, are also Tab crazy), but I sure have a dickens of a time finding it.

My cousin likes it because of its unique taste: cola with an afterburn. I personally hate that taste. It brings back memories of diets. That unique taste is the sweetener saccharin. Tab is the only remaining saccharin diet drink. Everyone else has switched to NutraSweet.

It seems as if they've been with us forever, but the fact is, diet soft drinks have only been available nationally for about thirty years.

The first attempt at a diet bottle was No-Cal, introduced in March 1952 by Kirsch Beverages of Brooklyn. It used cyclamate, a recently developed artificial sweetener, and it went nowhere.

The diet pop market got its first major player in 1962 when Royal Crown introduced its low-calorie soda pop, Diet Rite. But Royal Crown made a major marketing mistake: aiming it at fat people. No one wanted to drink something that was aimed at fat people. It was left to Coca-Cola to introduce a diet drink for people who were "watching their weight."

Tab made its debut in 1963 with the ad slogan: "Keep tabs on your calories with Tab." Even though it reminded many of us of a cat's nickname, Coca-Cola made a big deal out of the pains the company had gone to, to come up with the name. They'd even employed a computer! Some sixties computer nerd programmed the machine to list every four-letter English word. Coca-Cola wanted a short, catchy name, like, uh, Coke.

The computer wheezed and spit out a list of a quarter million words. Good job, computer. I have dictionaries with fewer words than that. My computer spell-checker has fewer words than that.

Among the less-than-appetizing possibilities were such wonderful four-letter combinations as Burp and Gaag and Flug. The company picked Tabb and then dropped the second *b*.

The ads touted the fact that Tab had only one calorie. That put it down there with water as a low-calorie drink. That's pretty much what it was: colored, carbonated water sweetened with that old standby artificial sweetener, saccharin.

Tab was an instant success. So successful that Pepsi had to race to get Diet Pepsi on the market.

Tab and Diet Pepsi fought tooth and nail for the diet pop drinkers and by 1982 Tab led with 4 percent of the total soft drink market. That's when Diet Coke was introduced and it defined instant success. Coca-Cola claimed it had finally developed a diet drink that was deserving of the name Coke, and shoppers bought, in record numbers.

And Tab became the ugly stepsister.

By 1987 Coca-Cola had even stopped advertising Tab.

Today one of every five soft drinks sold is a diet cola. But Tab is only a small part of that; it has only a 0.3 percent share of the soft drink market. New Coke outsells it five to one.

Coca-Cola says there are 5 million loyal Tab drinkers. And they have to be loyal. It's not in Winn-Dixie today. Scratch that off the grocery list. We'll get two Diet Pepsis instead.

2-LITER BOTTLE DIET PEPSI—$1.29

We buy regular Coke and Diet Pepsi because Judy thinks Diet Pepsi is better than Diet Coke. I can't tell the difference. One glass of cola-flavored NutraSweet water is the same as the other to me. She and I drink the diet soda; the theory is the calories we save here can be applied elsewhere. Except one look at my waistline indicates it isn't working.

I've been a diet soft drink consumer since Diet Pepsi was introduced back in the sixties. Those first Diet Pepsis tasted horrible. But they were low-cal, thanks to the substitution of the artificial sweetener cyclamate for most of the sugar.

Substitution is something the food industry has always had an interest in. Sometimes substitutes have been used for health and diet reasons. Other times it's been strictly for monetary gain. The first food substitute was a sugar substitute, saccharin, which was accidentally discovered in 1879 by scientist Constantine

Fahlberg. We can thank Fahlberg's poor hygiene for the discovery. Fahlberg, who had been working in his lab, neglected to wash his hands after he concluded. Later he touched his lips and tasted something sweet. It was 2,3-dihydro-3-oxobenziososul-fonazle: saccharin, a substance 500 times sweeter than sugar. In 1886 the *National Beverage Gazette* reported, "Saccharin is so sweet that a teaspoonful converts a barrel of water to syrup." Some food producers began using it then, to save money.

Saccharin was first challenged by Bureau of Chemistry head Harvey Wiley in 1907 and was banned in food but—oddly enough—not in chewing tobacco. The ban was lifted during World War I and saccharin was often used in combination with sugar. The 1938 federal food and drug law allowed it to be used in foods as long as the product had an appropriate labeling.

When food laws were rewritten with the Food and Drug Act of 1958, saccharin was placed on the "generally recognized as safe" list. But challenges continued and it was taken off the list in 1972. Now any product containing saccharin must have a label that tells buyers it contains saccharin, a "non-nutritive artificial sweetener for persons who must restrict their intake of ordinary sweets."

FORKLORE #28

SUGAR, INC.

The Coca-Cola Company buys more sugar than any other company in the world. Coca-Cola also buys more vanilla than any other company in the world. Madagascar is the world's largest supplier of vanilla and when Coca-Cola dumped the original Coke for New Coke in 1985, the economy of Madagascar went into a slump because New Coke contained no vanilla. Fortunately for Madagascar, Americans rejected New Coke, and the original Coke, renamed Coca-Cola Classic, returned to the grocery shelves. And Madagascar's economy returned to normal.

In the meantime, scientist Michael Sveda, another researcher with poor hygiene habits, had accidentally discovered a new sweetener. Sveda, who failed to wash his hands after work one day in 1937, was smoking a cigarette when he detected a sweet taste. He traced it to aspartylphenylanlaninine: cyclamate, a sweetener thirty times sweeter than sugar. With saccharin yo-yoing between approved, then banned, then approved, then banned, then approved with restrictions, cyclamate moved in. It offered one improvement over saccharin. Saccharin had a faint aftertaste of bitter almonds. Cyclamate had no aftertaste.

The first widespread use of cyclamate was in the diet soft drink Diet Rite, which was introduced in 1962. Within a year it was the fourth best-selling soft drink. Cyclamate consumption boomed, from 250,000 pounds in 1955 to 17 million pounds in 1969. By October 18, 1969, 175 million Americans were ingesting cyclamate in some form.

That day happened to be the day that the Food and Drug Administration banned cyclamate from the food supply. Ten days earlier an Abbott Laboratories scientist, working at the Food and Drug Research Laboratories on Long Island, discovered bladder tumors in rats who ate a diet composed of 5 percent cyclamate. It was an enormous amount of cyclamate, the equivalent of drinking 550 Diet Rites a day. A Coca-Cola official at the time noted, "You'd drown before you'd get cancer." But tumors were tumors.

So cyclamate, the heart of the first diet soft drinks, was banned. And saccharin made a comeback in diet foods. Then saccharin was once again on the hit list: It was temporarily banned by the Delaney Act of March 9, 1977. The ban produced a rush on supermarkets as people stocked up on Tab and Sweet'n Low. The ban was soon dropped, but another artificial sweetener was on the horizon.

Aspartame was discovered accidentally in 1965 at G. D. Searle and Company labs during research on ulcer drugs. It wasn't technically an artificial sweetener because it was a combination of two naturally occurring amino acids. It was 200 times sweeter than sugar. Searle sought FDA approval for its use in dry foods and as a table sweetener in 1973, and it was approved in 1974. But challenges kept it off the market for other uses until 1981, when it was approved for use in dry foods. The big breakthrough came in 1983, with aspartame's approval for use in carbonated beverages. Today 70 percent of the aspartame sold is used in soft drinks. You know it; it's sold under the brand name NutraSweet.

NutraSweet has been a sensation, bringing in $800 million a year for its parent company, Monsanto.

But it has created problems in the baking industry. Because it is so concentrated, not nearly as much is needed to produce the same sweet taste. A half cup of sugar can be replaced by an

---

FORKLORE #29

MR. WIZARD MEETS BETTY CROCKER, PART I

Perhaps you've heard the story. It goes something like this: At Harvard they left a fly in a Coke overnight and the next morning, the fly had been completely dissolved. The name of the university changes and so does the item to be soaked overnight, but the result is always the same: Coke eats it. The lesson is that if it does that to a fly, just think of what it does to your stomach.

To test this theory I swatted two flies: a test fly and a control fly. I put the test fly in a cup of Coke and let it soak for twenty-four hours. I put the control fly in a cup of Roto-Rooter drain cleaner and let it soak an equal length of time.

When I returned to the Coke fly the next day, I discovered, to my surprise, the fly floating around, unscathed. The Roto-Rooter fly, on the other hand, was dissolved down to a couple of tiny fly bits. The Roto-Rooter had also eaten through the bottom of the plastic cup.

I think there are two lessons here: Don't believe all those Coke stories you hear. And don't, for any reason, let a fly drink Roto-Rooter.

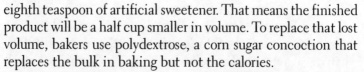

eighth teaspoon of artificial sweetener. That means the finished product will be a half cup smaller in volume. To replace that lost volume, bakers use polydextrose, a corn sugar concoction that replaces the bulk in baking but not the calories.

Sugar substitutes are an accepted fact in the American diet. NutraSweet is used in everything from Diet Coke to Kool-Aid to Jell-O.

2-LITER BOTTLE 7UP—$1.49

In 1920 in Price's Branch, Missouri, C. L. Griggs invented an orange drink. He called it Howdy and folks in those parts took a liking to it. There was just one problem. Howdy was all he could call it. He couldn't call it an orange drink because then it would have to have orange juice in it. He tried to improve on it, adding a few different flavors. The result was a drink he called Bib-label Lithiated Lemon-Lime Soda. It had more carbonation than a standard soft drink and it also contained lithium. Great drink, terrible name. So he tried to come up with another name. After six tries he gave up and called it 7Up. (Today's formula leaves out the lithium.)

7Up was the number-three soft drink in the sixties behind Coke and Pepsi. Today it ranks eighth.

Potato Chips, Pretzels

▰▰▰▰▰▰▰▰▰▰▰▰▰▰▰▰▰▰▰▰▰▰▰▰

In this aisle it's potato chips as far as the eye can see: Mike-Sells Potato Chips, Keebler's O'Boisie Potato Snack Chips, Golden Flake Potato Chips, and then, the Frito-Lay potato chips: Ruffles and Bar-B-Q and Sour Cream & Onion. Fully half of all the potato chips are Frito-Lay.

Who decides which potato chips go where and how many of each to put on the racks?

Money decides. It's called the "slotting allowance" and it's the eighties invention that no one wants to talk about. In many grocery stores food manufacturers pay to get shelf space. The fight for shelf space has always been vicious. If a product can't get on the grocer's shelf, it can't get into your home. But it wasn't until the last decade that some supermarkets discovered they could turn even that into a profit center.

A 1990 Gallup Poll survey of Snack Food Association members found 51 percent reporting they had to pay slotting allowances to get and keep their products on the shelves. Some told Gallup that stores were charging as much as $250 per section foot per contract year.

Grocery stores claim they have to charge, that there are so many new products coming and going in the marketplace, so much stocking and pulling and restocking with a different item, that they have to charge the folks responsible for all the churn, the manufacturers who keep turning out new products.

How much money are manufacturers paying per store? How much can a store expect to earn from slotting allowances? No one will say. But Pat Collins, president of Ralph's, a 127-store chain in southern California, told *USA Today* that his company spent $20 million in 1990 adding 10,000 new products and dropping 7,000. He said slotting allowances covered about 60 percent of the cost. A little simple mathematics tells us that companies paid Ralph's about $12 million in slotting allowances. Divide that by 127 stores and you get $94,000 per store.

14.5 OZ. BAG LAY'S POTATO CHIPS—$2.79

The potato chip is king among salty snacks. In 1990 Americans ate 6.1 pounds of potato chips each. Each! There are four in our family, so we must have downed almost 25 pounds of potato chips last year. That's a "pounder" bag every other week.

If you count potato chips as a vegetable, then that's terrific.

Now, potato chips are as American as a food can get. They were invented in Saratoga Springs, New York, during the summer of 1853 by an American Indian named George Crum. Crum was the chef at the Moon Lake Lodge resort there. One evening a picky guest sent back his French fries because they were too thick. Crum patiently sliced another potato, just a little thinner. When the guest sent back this second batch, Crum decided to show him. He sliced the next round so thin you couldn't pick them up with a fork. The guest loved them and Crum was soon forced to add Saratoga Chips to the menu.

They don't hand slice potato chips anymore. They're machine

cut. Frito-Lay has thirty-five plants around the country that make potato chips. Those plants use 7 million pounds of potatoes a day.

They buy potatoes by the railroad car, potatoes from all over the country. All different. Potatoes have a natural balance of sugar and starch and Frito-Lay tests each batch of potatoes they buy for balance. If the potatoes deviate from the standard, the company cures them until they have the right balance.

They are then scrubbed and peeled by machine. A spinning chip cutter slices them. Straight blades produce regular potato chips; rippled blades, Ruffles.

12 OZ. BAG CHEE-TOS—$2.49

Chee-tos were invented by the Frito-Lay company in 1948; the name is an attempt to connect the cheese-flavored snack to the company's popular corn chip, Fritos.

Chee-tos are a parent's nightmare. That orange powder coating turns to wet orange cement when it comes in contact with a child's hands. Is there a family room couch anywhere in America that doesn't have a Chee-tos stain somewhere?

Look on the package and you'll notice the primary ingredient in Chee-tos isn't cheese. It's cornmeal. Cheese doesn't make its appearance until two ingredients later, after vegetable oil and whey.

Frito-Lay begins with stone-ground cornmeal, adding moisture to turn it into dough. The dough is then pushed through the holes of an "extruder," a fancy industry term for a die. When the hot Chee-tos dough meets the cool room temperature, it "explodes," sort of like popcorn. A knife then cuts the popped extruded pieces of dough into bite-size bits.

Chee-tos Crunchy, the ones you think of when you think of Chee-tos, are then "crisped" in vegetable oil (you might call it "fried"). Still no cheese, but we're almost there.

In the last step before the Chee-tos are dumped into the bag, they are dumped into a seasoning drum. Now, finally, the cheese is added. Artificially colored Cheddar cheese seasonings (that bright orange color is a combination of yellow number 6, turmeric, and annatto), salt, and vegetable oil are sprinkled on the Chee-tos. That's why that orange stuff gets all over your fingers when you eat them. It's not baked in; it's added afterward.

In 1990 Frito-Lay, makers of Chee-tos, Fritos, Doritos, and Tostitos, began what might be considered the height of fool-ishness: making junk food healthier. The company introduced Ruffles Light Choice Potato Chips in two flavors, regular and sour cream and onion; Doritos Light tortilla chips in two fla-vors, nacho cheese and cool ranch; and—yes!—Chee-tos Light. These new, healthier junk foods were ten years in development and cost Frito-Lay a whopping $150 million in start-up costs, according to press reports.

Why take such a huge chance? Because Americans spend $30 billion a year—$120 per person—on snack foods, $10 billion of that on salty snack foods.

How do you make junk food less junky? By using a third less oil. That means fifteen Doritos Light chips contain only 110 calories and 4 grams of fat, as opposed to fifteen regular Dori-tos chips, which have 140 calories and 7 grams of fat. Hey, that means you can eat four more Doritos Lights.

FORKLORE #30

MR. WIZARD MEETS BETTY CROCKER, PART II

Want to see what Chee-tos really look like? What to see the soul of a Chee-to? Then perform this experiment at home.

Take a Chee-to and place it under running water (lukewarm works best). Watch as the orange stuff washes right off. What's left is the popped extruded corn dough, a Chee-to without the Chee.

Set aside until dry. Taste it. It still has that crunch, but without that orange goop caked on your fingers, it's just another extruded corn snack.

---

7 OZ. CANISTER PRINGLES
POTATO CRISPS—$1.49

Pringles, the stackable potato chips (had America been asking for a potato chip that could be neatly stacked?), created a revolution. No, not because you could stack them. A revolution in the potato chip delivery system. Before Pringles most potato chips were local or regional brands. They had to be: Potato chips had a short shelf life. And the longer they were in transit, the more the chips got broken. Procter & Gamble solved both of those problems. Because the Potato Crisps were stacked in a tennis ball can, they didn't break as easily as bagged chips. And because the can was vacuum-packed, they stayed fresher longer. So Procter & Gamble could deliver Pringles through the same grocery warehouse system it used to deliver Tide detergent.

Procter & Gamble also solved another problem that had long burdened the potato chip industry: Whole potatoes are cumbersome to handle in production. P&G eliminated that problem by using chopped-up potato flakes, which were mixed with water, salt, sugar, and a little oil to form a dough. A die then stamped out cookie-shaped disks from the dough and the disks were fried in a curved screen that gave the chips their saddle shape.

Pringles was test-marketed in Evansville, Indiana, in the fall of 1968. It made its national debut in 1971, at the same time that General Foods introduced Pringles pretzel snack.

Both companies thought they had clear title to the name, but it turned out only one did: General Foods. So the two companies settled the matter out of court with Procter & Gamble paying General Foods an undisclosed sum for rights to the Pringles name.

That was just the first court battle involving Pringles. Companies that made "real" potato chips asked the Food and Drug Administration to force P&G to put the words "imitation potato chips" on the Pringles can because Pringles were made from dehydrated potatoes. To call itself a potato chip a product must be made from raw potatoes. The FDA ruled in Procter & Gamble's favor, saying Pringles could be called potato chips as long as it had the words "made from dried potatoes" on the package.

Of course, if Pringles had been a bomb, the other potato chip makers wouldn't have bothered. But Pringles was a hit—an instant one. By 1975 Pringles had about 10 percent of the potato chip market with sales of $110 million. But the brand had peaked and a few years later sales were dropping.

The researchers at Procter & Gamble discovered they had two problems: Shoppers perceived Pringles as expensive and they were bored with only one flavor. So in 1981 P&G brought Pringles out in single canister packs, instead of double canisters, and added six new flavors.

Pringles has made a comeback since then and is now the fifth most popular salty snack.

FORKLORE #31

WE INTERRUPT THIS BOOK . . .

The phone call came during dinner. Don't they always. My wife, Judy, answered and a pleasant-voiced lady asked her if she would participate in a market research study.

Judy agreed.

"Why did you say yes?" I wanted to know.

"Because I like to do those things," she said, defensively.

Her end of the deal sounded simple. All she had to do was watch an episode of *Get Smart* on our local independent TV station and answer a few questions when the woman called back the next evening.

"Why are they researching *Get Smart?*" she wondered out loud. "I thought that show was just reruns."

I kept silent. I knew the answer.

As evening turned to night and ten-thirty—*Get Smart* time—approached, Judy began yawning.

Finally, at about ten, she gave up. "Would you tape it?"

I said I would.

When I walked in the door the next afternoon, I asked Judy if she had watched the tape.

"Oh, I forgot all about it," she said. So I started the tape for her and went off to another part of the house to do another-part-of-the-house-type things.

I returned to find her skipping past the commercials.

"I forgot about soccer practice and PTA," she explained as she pressed the speed-search button. "I don't have time to sit through all the commercials."

I kept silent.

The pleasant-voiced lady called during dinner. Again. My wife was on the phone with her for half an hour.

I had given Judy only one instruction. I told her to ask the pleasant-voiced lady what it was all about. Any time you donate—and donate is the proper word—your time to any of these telephone research groups, they have an obligation to tell you what they are studying. That is an ethical standard of the major psychological research groups, but market research and psychological studies are not the same thing.

"Oh brother," Judy said when she returned from the phone to normal life.

I nodded. I knew why she was oh-brothering.

"She started by asking what I thought about the show, was it excellent, very good, good, fair, or poor. Then she asked if I watched it in its entirety. Then she asked if in the show there was a man who was supposed to die in Smart's bedroom, I guess to find out if I knew the main idea of the plot. Then the rest of the time she asked about the commercials. And I had speed-searched through most of the commercials," she moaned.

The pleasant-voiced woman had asked her if she remembered a commercial about corn chips (she did) and a commercial about tampons (she remembered Brenda Vaccaro) and a commercial about nasal spray (she didn't because she had skipped it).

Then she was asked if she remembered the brand names, the people in the commercials, and what ideas she thought the commercials were trying to get across.

She kind of tap-danced on these. She had seen most of the commercials before, but not from *Get Smart*.

At the end of the survey Judy asked the pleasant-voiced lady what it was all about and the woman told her they were studying commercials. Not that my wife hadn't gotten the idea after one minute of questions about the show and twenty-nine minutes of questions about the commercials.

The moral of this story is that if some woman (or man) calls you on the phone and asks you to watch a rerun of an old show because she wants to ask you a few questions, it isn't because she wants you to watch the old show and it isn't just a few questions.

Health Foods,
Gourmet Foods,
Diet Foods,
Imported Foods

▲▼▲▼▲▼▲▼▲▼▲▼▲▼▲▼▲▼▲▼▲▼▲▼▲▼▲

Judy almost never shops this aisle. And it's not because she isn't health conscious. She's like a charter subscriber to *Prevention* magazine. She has one simple reason for never shopping this aisle. And I'll let her say it: "Too expensive."

If you watch TV or read the newspaper, you'd think that the health food craze is sweeping the supermarket shelves. The truth is, in the grand scheme of things hardly anyone buys health foods. But it is the fastest-growing segment of the grocery bill and the increased awareness of health is being felt in other aisles as well. Froot Loops and Fruity Pebbles and Smurf-Berry Crunch have been joined in the cereal aisle by such healthy-sounding products as Müeslix, Stone Ground Wheat Flakes, High Fiber Farina, and 4-Grain Cereal with Flaxseed.

The fact is, we Americans talk a good game, but when it comes to putting our food dollars where our mouths are, healthy eating loses. In fact, fewer people served fresh foods as part of a main meal in 1989 than in 1985. When asked why they didn't cook fresh foods more often, respondents complained that fresh-food preparation was too time consuming.

Well, whine.

How much are we changing our eating habits?

In 1980 the four main meal entrées in surveyed households were:

1. Ham sandwich
2. Hot dog
3. Steak
4. Cheese sandwich

Nine years later they were:

1. Ham sandwich
2. Hot dog
3. Chicken
4. Steak

FORKLORE #32

I'M IN THE MOOD FOR FOOD

Do our moods send us running for certain foods? Two researchers at Simon Fraser University in Canada tested that thesis in 1988, using students, that old research reliable, as subjects.

The results are in.

We prefer crunchy foods when we are feeling amused or bored; sweet foods when we are feeling worried, amused, or bored; spicy foods when we are feeling friendly, happy, confident, or amused; salty foods when we are feeling anxious, amused, or bored; sour foods when we are feeling sad; warm foods when we are feeling confident, friendly, happy, relaxed, or solemn; and chilled foods when we are feeling happy or relaxed.

And when we're in the mood for love? We want chilled, spicy, or sour foods.

2 LB. BAG JOLLY TIME POPCORN—$1.49

Okay, I'm sort of ashamed to add this to the cart. That name: Jolly Time. Where's Bozo and Cookie?

But Jolly Time and Orville Redenbacher's are the only pop-your-own popcorns on the shelf, and since I'm not in the mood to pay three dollars for Orville's stuff, I'm stuck with Jolly Time. The rest of the popcorn space in the supermarket has been given over to microwave popcorn.

I can still remember my words the first time I saw microwave popcorn back in 1979: "How lazy can you get?" I really didn't think microwave popcorn would ever catch on. Ten years later it had knocked pop-your-own popcorn into the gourmet aisle.

We switched to microwave popcorn about five years ago. It was more convenient. But about a year ago we switched back. Pop-your-own tastes better.

I prefer Jolly Time over Orville Redenbacher's because it pops bigger. Also, I hate Orville Redenbacher's commercials.

Jolly Time, which is manufactured by American Pop Corn Company, of Sioux City, Iowa, is the oldest brand name in popcorn. It has been around since 1914 and was the first popcorn sold in retail stores. It was originally sold in glass jars, but in 1925 the company switched to less expensive cans. In 1957 Jolly Time changed to two-pound plastic bags and that's been the industry standard ever since.

But the finest contribution Jolly Time has made to the world may be the nonexploding popcorn bag. This paper sack, designed for use in movie theaters, can't be inflated and then exploded by smacking it with your hand. No more loud bangs in the middle of a Hitchcock thriller.

Milk, Dairy Products, Bread

▲▼▲▼▲▼▲▼▲▼▲▼▲▼▲▼▲▼▲▼▲▼▲▼▲▼▲▼

More than any other section of the food market, the dairy case is dominated by store brands. Next to Shedd's margarine, in a tub that's the same color with similar lettering in the same place on the package, is Superbrand margarine. Kraft American Cheese Singles is flanked by the near-clone Superbrand American Cheese Singles. Dean's 2 Percent Milk is surrounded by Superbrand 2 Percent Milk.

Superbrand is Winn-Dixie's store brand. There are some products Judy would never buy in store brands. Milk is not one of them. We have always bought Superbrand milk.

Store brands used to have a bad reputation. My mother thought they were the hind end of a company's production line, the stuff that wasn't good enough to carry the Nabisco label or the Kraft label and was sold off to packers who turned them into store brands.

Store brands are always cheaper than national brands and that has given them the reputation as the brands of poor people. But that isn't so. Everyone buys some of them, no matter whether they drive them home in a Mercedes or a Yugo.

Store brands date back to the nineteenth century, when A&P introduced Eight O'Clock Coffee and A&P Baking Powder. Originally they were introduced for the low end of the price and quality scale. But customers rejected them.

They are now frequently as good as name brands. President's Choice, a store brand sold in Bell's and D'Agostino's supermarkets, won in two categories in the 1990 Retail New Products Contest. Sometimes, but not always, the store brands are the same as the national brands, with a different label. A number of large companies, including Land O'Lakes, the dairy firm, package a part of their production runs in store brand containers.

Supermarkets like store brands: The profit margins are higher than for name brand items. Stock clerks are told to position the store brands to the right of national brands. Sure enough, every store brand in the dairy case is on the right side of its national competition. That's because more people are right-handed and they will have to reach across the store brand to get the higher-priced national brand. Does that discriminate against left-handed shoppers?

I GALLON SUPERBRAND
I PERCENT MILK—$2.19

When I was a kid, we would occasionally spend the night at my grandmother's farm. And the next morning my aunt Nola would go out early and milk the cow, strain the milk through an old dish towel, and pour it into our glasses and over our cereal.

That was great milk, so full and tasty. That was real milk.

We don't drink real milk at our house anymore. We drink 1 percent milk. That means that 1 percent of the milk is butterfat and the rest is water and milk proteins. And it tastes like it.

I have never acquired a taste for it. As it turns out, I have developed an allergy to milk, so I don't have to drink it anyway.

But my wife drinks it and my sons drink it. And it's a lot healthier, we are told.

But for someone who's tasted real unpasteurized milk straight from the cow's udder, 1 percent milk tastes like thick white water.

Our family is not the only one to switch from whole milk. In 1970 Americans averaged drinking 202.9 pounds of whole milk per person a year. In about twenty years that had dropped by half. (The dairy industry insists on measuring milk in pounds instead of gallons. One pound equals 2.1 quarts.)

In addition to having less butterfat, virtually all modern milk is pasteurized. Essentially that means that milk is heated to a temperature of about 160° F., to kill bacteria, then cooled to 50° F. We've all heard the story of Louis Pasteur and how he saved France and the world from tuberculosis by figuring out the process of pasteurization. Actually, it was Pasteur's associate, the chemist Andre Soxhlet, who applied the pasteurization process to milk. In 1886 Soxhlet recommended that all milk being fed to infants be heated first.

And very few people know the story of Nathan Straub, whose contribution to pasteurization was almost as important as Pasteur's. Straub shoved pasteurization down a reluctant populace's throat. Straub, a New York philanthropist, set up his own milk-processing stations in the early years of this century, at a time when many of New York's children were dying from TB before reaching the age of six. At his urging New York City created the post of inspector of dairy farms in 1906, making it one of the first cities to check on its milk at its source. The first had been neighboring Newark, which had passed a law in 1882 providing for inspection of dairies.

At the turn of the century some dairies were already pasteurizing their milk because they knew it would retard souring and increase shelf life, but they were doing it in secret because not everyone agreed with Nathan Straub. The pasteurization movement got a big push in 1907 when President Theodore Roosevelt's milk commission formulated three rules for healthy milk: cleanliness, cold and speedy transportation, and pasteurization. But it was Chicago, not New York, that first adopted compulsory pasteurization; it did so in 1909 and soon pasteurization was the law everywhere.

Pasteurization had a dramatic effect on infant deaths. In 1885 the infant mortality rate in New York City was 273 per 1,000 live births. By 1915 it had fallen to 94 per 1,000.

Pasteurization was one thing: People accepted that it provided health benefits. But homogenization—the breaking up of the fat globules to make the milk creamier—was another. It was a tough sell to milk drinkers used to that familiar cream line at the top of the milk bottle.

But a unique product requires a unique sales approach and that's what homogenized milk got in 1932 from McDonald Dairy of Flint, Michigan. In order to convince skeptical customers that homogenized milk was better for the digestive system, the dairy hired a panel of men. Half drank regular milk, the other half homogenized milk. After a period of time the men would vomit the milk back up and the dairymen would show interested customers—if there were any who wanted to pick through vomit—that the curd had been digested better in the group that had consumed homogenized milk. The curd samples were then preserved in formaldehyde in glass jars and route salesmen carried these jars around to prove to their customers that homogenized was best.

That's right, milk salesmen were carrying around jars of vomit. I can hear the sales call now: "Uh, excuse me, ma'am, uh, I have right here in this jar. . . ." How many doors got slammed in those guys' faces?

The same year that McDonald Dairy sold the public on homogenization, the push for vitamin D–fortified milk began with the publication of an editorial in the *Journal of the American Medical Association* urging that vitamin D be added to milk. A decade earlier Dr. E. V. McCollum of Johns Hopkins University had discovered that vitamin D could prevent rickets, a bone disease prevalent among children at the time. Late in 1932 Borden offered the first commercially available vitamin D milk in Detroit.

Pasteurization, homogenization, and fortification aren't the only processing that happens to milk. There's another process that should be called homogenization, since it reduces all milk to a common denominator. That's the process of dumping all farm milk together at the dairy. If one farmer has bred a cow for her tasty milk, it all comes out at the dairy.

Milk was not a grocery store item until after World War II. The first supermarkets carried milk only as a service to customers who might have discovered they needed an extra quart after the milkman had passed. In 1945, 80 percent of retail milk sales were through home delivery. Today almost none is.

The war had a lot to do with the switch to grocery store milk. Before the war most city dwellers were used to every-day delivery by the milkman. During the war that was cut to every-other-day delivery to conserve gas, and people began to develop the habit of buying milk at the grocery.

The milk bottle on the back doorstep was a staple of life in the thirties, forties, and early fifties. But the dairy industry was never enamored of the old-fashioned milk bottle. Bottles were a hassle. Bottle washing was a large expense for the dairies. And invariably some bottles were broken, lost, or stolen. In 1937 it was estimated that lost bottles cost dairies $12.5 million a year.

There was already a movement to find a cheaper, more trouble-free replacement. One of the first attempts was the paper bottle invented in 1906 by G. W. Maxwell of Los Angeles. He couldn't get any of the local dairies interested and his idea died.

The first square paper carton, the now-familiar Pure-Pak, was developed at Ohio State University in 1926. It was not an immediate hit because the machine that manufactured it required three operators and produced only ten cartons an hour. But scientists continued working on the machinery and in 1935 the Ex-Cell-O Corporation bought the rights to the process and began commercial manufacture.

The first Pure-Pak had to be cut off at the top; the familiar built-in spout was added in 1952. A perforated tab was developed in 1941 and that made the carton more attractive. In 1942 virtually all milk came in glass bottles. By 1967, 70 percent of all milk sold was in paper cartons. That worked just fine in the quart size. But the half gallon was unwieldy. And the gallon carton was bulky and difficult to pour from. So dairy operators continued their search for a better container. And a bigger container, to encourage consumption. Hence the one-gallon plastic jug.

20 OZ. LOAF BUTTERNUT ENRICHED BREAD—$1.19

Today it's called white bread. We called it light bread when I was growing up, to distinguish it from dark bread, which didn't need any distinguishing anyway, since nobody bought it. Light bread is one of those technological innovations that isn't necessary. By extracting a third of the wheat kernel's bulk—the bran and the germ—bakers can produce a bread that is almost snow white on the inside. But gone with the bran and the germ are many of the nutrients.

White bread took the country by storm in the 1840s. People liked its looks. It looked purer. There was also a snobbery associated with white bread. Better people ate white bread. Snobbery in the matter of bread color was nothing new. It dated back to the Romans. Roman senators and army officers took pride in offering their guests white bread. Dark bread was for the working classes.

White bread consumption in this country increased every year until 1963. Americans consumed 9 billion pounds that year and that total has been declining slowly ever since. In 1982 it was 6.2 billion pounds. By 1990 it was only 5.6 billion pounds.

We all know that white bread has many of the nutrients and much of the fiber removed in processing. We all know that whole grain breads are better for us. From all the health articles on the subject, you'd think the whole country had switched back to whole wheat bread. Think again. The decrease in white bread sales are part of an overall decrease in bread consumption. White bread is still *the* bread. Sales in 1990 totaled $5.1 billion. Variety breads, that's all the weird stuff—whole wheats, ryes, pumpernickels—accounted for only $970 million in sales. White bread still outsells whole grain bread by more than five to one.

Before 1930 most bread was sold in whole loaves. Housewives sliced it at home. But that year the "sandwich loaf" was born and with it the cliché "that's the best thing since sliced bread."

Bread machines that take the drudgery and the guesswork out of baking bread are popular in upscale households, but most people still buy their bread from someone else, usually a supermarket.

Most bread is used in sandwiches. The next most popular way to eat it is to toast it. The Egyptians were the first to toast bread, probably around 2500 B.C., but not because they were looking for a side dish for scrambled eggs. It was a method of preservation. Those first toasters were nothing more than a couple of long-handled forks. It would be four millennia before the next improvement in toasting equipment.

It was called a toaster oven, but it was nothing like the ones we have. It was really just a wire cage that angled a piece of bread above the burner on the top of a wood stove. It was popular in the nineteenth century. Electric toasters arrived in homes shortly after the turn of the century. It was a quantum leap in toaster technology: It meant you didn't have to start a fire just to have toast. But it was far from perfect. It took a while to heat up and when it did, it burned too hot. The first piece of toast was usually undercooked; piece number two was perfect; everything after that came out burned.

It was left to a factory worker in Stillwater, Minnesota, to perfect the toaster. Charles Strite, who was tired of the burned toast his company's cafeteria served, fiddled around with some springs and a timer until he had it: the pop-up toaster. His patent was granted on May 29, 1919.

The first one hundred were hand made by Strite in his garage and sold to the Childs restaurant chain. Within two months every one of them was back on Strite's doorstep. A spring had broken on one; the timer quit on another. But Childs officials didn't want a refund. They loved the idea of a toaster that popped out the toast rather than burned it. So Strite went back to the drawing board until he had the perfect combination of springs and wires.

The first home pop-up toaster, the Toastmaster, was introduced in department stores in 1926. It was the ultimate toasting machine. It even had a darkness timer! A ready public went wild. Sales skyrocketed and Congress even got in on the mania, declaring March 1927 as National Toaster Month.

My grandmother used to make her own bread: she'd knead the dough until it was just right, then let the bread rise and bake it. Anyone who has ever tried to make bread at home—in the days before bread machines—knows how truly difficult it is.

It isn't any easier for the big bread companies, some of which turn out 20,000 loaves an hour.

A bread plant can go through 2 million pounds of flour a week. When you use flour in that kind of quantity, you don't buy it in 5-pound bags. Bread plants order flour by the ton and have it delivered in special steam-sterilized freight cars or tanker trucks.

A large commercial bakery stores flour the way farmers store feed, in large silos that hold upwards of 150,000 pounds each. The flour in each silo is marked by its arrival date and—just as the home cook does—the oldest is used first. Because there is so much flour around, bakers have to be very careful. The one sacred rule at all bread plants is no smoking: One careless match can turn a bread factory into a crouton factory.

The first variable in bread baking is the flour. Because no two wheat crops are alike, no two flours are exactly alike. One may have more moisture, another more starch. So bakers mix the flour. They check for moisture absorption and adjust recipes accordingly.

The next variable is the water. Because the hardness can vary from the morning to the afternoon, bread plants have to keep a constant check on the water. Some even test it every hour.

The ingredient in bread that causes the most difficulty is the yeast, which is what makes the bread rise. Yeast is very sensitive to temperature and humidity. When the weather is warm and the air is moist, the yeast works too rapidly and the bread rises too fast. Bakers love driving to work on cool, dry days because they know the yeast will act like it is supposed to.

Bakeries measure their ingredients by weight and load them into bread mixers as big as cement trucks with paddles the size of boat oars.

The machine does the rest, mixing the dough into a tight ball and dumping it out for a rest. This is where bakers become like officials at a track meet. Their timing must be perfect. The rest can't be too short and it can't be too long. If it's too short, the top of the loaf will separate from the bottom during baking. If it's too long, the bread will get too big for the pan.

After its nap, the dough is put through a machine that punches the gas out with a row of giant rods. When the dough has belched out all of its air, it is cut into loaves. The loaves are then dropped into a machine called a rounder, where they are tumbled and dusted with flour. The dough is now ready for its final rest, a nap of about ten minutes.

The next machine on the line squashes each loaf flat, rolls it up, seals the seam on a pressure board, and plops the loaf into a pan. The seam of the loaf must be on the bottom. The pans move into a temperature-controlled box, where they stay for about an hour. When the pans come out, the dough has risen; it is puffed up and ready for the oven.

The oven isn't like the one at Pizza Hut, with one baker shoving the pan in and another pulling it out when it's done. Bread is baked on a conveyorlike oven shelf and when it exits the oven at the other end, it should be done. There is it met by a row of suction cups that lift each loaf out of the pan and gently deposit it onto another moving line. This time it's into the cooler. When the temperature of the baked bread has fallen to 100° F., it can be sliced without coming apart. The slicing is done by a series of spinning blades that look like circular saws. As each sliced loaf reaches the end of the line, a puff of air blows open a plastic bag and the loaf slides in.

The packaged loaves are loaded immediately into boxes and delivered to supermarkets in the early morning hours. Once in the market, the bread usually stays on the shelf for two or three days. If no one buys it in that time, it's sent to the company's thrift store or to a feed company to be used for chicken feed.

Each bakery has a quality-control panel that periodically checks the bread's crust color, slicing, grain, crumb color, symmetry, and, most important, flavor and taste. The flavor is rated either delicate, neutral, gassy, musty, rancid, or off. The taste is labeled either pleasing, bland, sweet, sour, metallic, or gummy.

16 OZ. LOAF RAINBO LIGHT WHITE—$1.49

This loaf of Rainbo is baked locally for Campbell Taggart of Dallas, the second largest bread company in the country. It's called Rainbo here and in thirty other cities, Colonial Bread in 29 cities, Kilpatrick's Bread in two cities, and Manor Bread in one city.

The name Rainbo Light Wheat sounds disgustingly healthy until you read the ingredients: water, wheat flour, oat fiber, whole wheat flour, wheat gluten, corn syrup, less than 2 percent of the following: yeast, molasses, salt, wheat bran, dough conditioners (may contain one or more of the following: mono- and diglycerides, ethoxylated mono- and diglycerides, calcium and sodium stearoyl lactylates, potassium bromate, calcium peroxide, calcium carbonate), partially hydrogenated soybean oil, soy flour, yeast nutrients (monocalcium phosphate, calcium sulfate, ammonium sulfate), distilled vinegar, cornstarch, malted barley flour, calcium propionate added to retard spoilage, niacin, iron, thiamine mononitrate (vitamin B_1), riboflavin (vitamin B_2).

What does all this mean? It means Rainbo Light Wheat isn't as healthy as it sounds.

Let's start with "dough conditioners." My grandmother never used dough conditioners in her bread. I think she would have had a difficult time locating any calcium or sodium stearoyl lactylates anyway. But she also didn't bake enough bread to feed a city. To make dough tough enough to endure the machine baking process, the batting around and tossing about by giant paddles and tumblers, bakers add any number of these chemical additives. What they do is make the dough drier and more elastic. One of the dough conditioners, calcium carbonate, is also used as a gastric antacid and an antidiarrheal medicine.

Another suspicious-sounding ingredient is "calcium propionate added to retard spoilage." That's been an age-old problem with bread. It starts to get stale the very second it comes out of the oven. As bread ages, the starch crystallizes and makes the bread hard. Then mold begins to form. Calcium propionate— a preservative—slows down that process. Calcium propionate has also been used as an antifungal medicine on the skin.

It doesn't mention it on the label, but the flours were chemically matured. In the old days bakers had to let flour age, or mature, to oxidize the proteins; otherwise the dough would be a sticky mess that wouldn't rise. Now millers age the proteins with chemicals called maturing agents, such as acetone peroxide, chlorine dioxide, and potassium bromate or iodate.

FORKLORE #33

HEY, IT WORKS FOR BEAVERS. . .

In 1979 the Federal Trade Commission obtained a consent order against ITT Continental for claiming its Fresh Horizons bread contained five times as much fiber as whole wheat bread. It was true, the bread did have five times as much fiber, but that extra fiber came from wood, which the FTC said was "an ingredient not commonly used, nor anticipated by consumers to be commonly used, in bread."

We are so overloaded with nutrition information today that it's hard to believe there was a time when no one talked about vitamins and fiber and Recommended Daily Allowances. Vitamins A and B weren't even discovered until shortly before World War I. The Food and Drug Administration advised Americans that vitamins A and B were "necessary for health and growth as the better known constituents of food." A third vitamin, the antiscorbutic vitamin C, was discovered in 1918. And in 1922 scientists found the antiarthritic vitamin D.

The Committee on Foods and Nutrition, a scientific group impaneled by the National Science Foundation in 1940, issued a report the next year outlining Recommended Daily Allowances for each vitamin. The committee chose its words carefully, using "allowance" rather than "requirement" to allow for scientific advances. The committee also endorsed the recommendation of the Council on Foods of the American Medical Association, which in 1939 had suggested that the nutrition removed in processing bread, flour, and milk be restored. With the support of the American Bakers Association and the Millers' National Federation, bread and flour "enriched" with vitamin B complex, thiamine, nicotinic acid, and iron—essential nutrients that weren't then in the American diet—appeared in stores in March 1941. The industry preferred the word "enriched" rather than "restored." No one thought it ironic that bread had to have the nutrients removed in processing added back in.

Today bakers also add in calcium. Most nutritionists agree that the enrichment of bread has contributed to significant declines in a number of diseases including pellagra. I have to admit I've never known anyone with pellagra.

20 OZ. LOAF ARNOLD HONEY WHEAT BERRY BREAD—$2.29

Arnold Honey Wheat Berry Bread is made almost completely with white flour and gets its dark "healthful" color from raisin syrup. We're not buying it. I just wanted to point that out.

6-PACK SEALTEST LIGHT N' LIVELY
STRAWBERRY LOWFAT YOGURT—$2.38

I never ate a single helping of yogurt when I was a kid. Probably no kids did in the fifties. But I had heard about yogurt. It had a reputation as the worst food in existence.

Yes, we're buying yogurt today. But this yogurt is a far cry from the stuff of my youth. This yogurt tastes good. Or at least the fruit and the sweeteners that are in it taste good.

For this we can thank Daniel Carasso, founder of Dannon Yogurt. He added strawberry preserves to his product in 1947 and yogurt hasn't been the same since.

Still, yogurt sales didn't even register on the government's yogurtometer until Bengamin Gayelord Hausser's book *Look Younger, Live Longer* was excerpted in the October 1950 issue of *Reader's Digest.* Hausser exclaimed: "To add years to your life get acquainted with and use every day these five wonder foods: powdered brewers' yeast, powdered skim milk, yogurt, wheat germ and blackstrap molasses. . . . [Yogurt] is a 'must' on the 'Live Longer' diet. Among the Bulgarians, where yogurt is a part of each meal, but where diet is not outstanding in other respects, the life span is longer than that of any other peoples in the world; Bulgarians are credited with retaining the characteristics of youth to an extremely advanced age."

(Comedian Jimmy Durante would quip that Hausser's regimen didn't make you live longer, it only seemed longer.)

Yogurt sales climbed and it's easy to see why. Who didn't want to look younger and live longer? Of course, an increase in sales of yogurt was just a blip on the food chart of life. There were no yogurt sales to speak of before Hauser's book. Even soaring sales didn't put yogurt in the league with, say, jelly donuts. The yogurt makers would like that to change. They now feature men and kids in their ads—not just women. And recently I've noticed free Sealtest Light n' Lively Yogurt coupons on Diet Pepsi.

It isn't working. My wife is the one buying today's yogurt.

2 LB. BOX VELVEETA CHEESE SPREAD—$3.99

When it comes to twentieth-century American plastic food, this is ground zero. This is the one all others are compared to: A cheese that isn't really cheese. It's cheese food. Cheeselike food.

Not that Kraft has ever tried to hide the fact that Velveeta was processed cheese. On the box, right from the start, it said, "Pasteurized Process Cheese Spread." It's just that sixty-five years ago, when Velveeta was invented, the word *processed* wasn't a bad word. In fact, Velveeta was on the cutting edge of the American technological revolution. It was the first food product extensively studied by university researchers. Kraft set up a special research fellowship at Rutgers University specifically for the study of Velveeta.

"Hi, what did you major in, in college?"

"Oh, I majored in Velveeta cheese."

These researchers studied its digestibility and nutrient content. As a matter of fact, much of the research into riboflavin (vitamin B_2) originated with Velveeta.

Velveeta was originally developed in 1915 by Elmer E. Eldredge, a Cornell-trained bacteriologist and chemist who had been hired by the Phenix Cheese Company to duplicate Gerber's Swiss Gruyère cheese, a processed Swiss cheese. It was Eldredge's solution to the whey disposal problem in cheese making. Whey was the liquid that was allowed to run off the cheese. He took this waste product, separated the whey protein, then mixed and heated it with cheese and two ounces of sodium citrate, an antioxidant. He called it Phen-ett and if that name had stuck, we probably wouldn't be eating processed cheese today. Cheese food sounds bad enough. Phen-ett sounds downright medicinal.

Shortly after Eldredge developed his cheese food, scientists at J. L. Kraft developed a similar processed cheese and called it NuKraft.

Here's where pure dumb luck played a role in the invention

of Velveeta. In 1915 Eldredge filed for a patent for his processed cheese using American, Swiss, and Camembert cheeses. But the patent examiner insisted only one type of cheese could be patented. During the ten-month dispute Kraft filed its own patent and it sneaked in.

Phenix and Kraft later agreed to share the patent rights and Kraft renamed NuKraft—since it wasn't really new—with a name that gave shoppers an idea what it was like, velvety cheese: Velveeta. Velveeta arrived on grocery shelves in 1928, just in time for the Depression and the invention of the supermarket. The natural cheese makers howled, calling processed cheese "embalmed" and "imitation." But shoppers pronounced it good and made it a grocery store hit.

To understand processed cheese, you need to know about natural cheese. Cheese is made from the compressed curd of a cow's milk. A few years ago food scientist John Bedrook, director of science at DNA Plant Technology Corporation, joked, "I'm glad cheese was invented nine thousand years ago, because today if someone suggested throwing milk and 'bugs' together to make something to eat, they'd get laughed out of town." But in essence that's how natural cheese is made.

Lactic acid bacteria are added to fresh milk and the mixture is heated to 86° F. Rennet extract, which is obtained from the inner lining of the fourth stomach of a calf, is added, setting up coagulation, the separation of the curds—the milk solids—from the whey—the milk liquid. The whey is drained off, leaving the semi-solid curd. It is heated to about 100° F., to expel the moisture, then salted and put into molds. The molds are taken to a curing room and kept there at low temperature for anywhere from a few weeks to several months to ripen. It takes about five quarts of milk to make one pound of cheese.

Processed cheese takes the shortcut. It's made from the scraps of different natural cheeses and the odds and ends of unripened cheese, then pasteurized and blended with seasonings or texturizers and sold without ripening.

American cheese is processed cheese made from any combination of Cheddar cheese, washed curd cheese, Colby cheese, and granular cheese (uncompressed curd with a granular texture). By law it must contain at least 45 percent butterfat and not more than 44 percent water. It may also contain an emulsifier to keep the fat and the water mixed, an acidifying agent for tartness, cream, mold inhibitors, lecithin to prevent prepackaged slices from sticking to the plastic wrappers, salt, and artificial color.

The definition of cheese food is less restrictive. It must contain not less than 23 percent butterfat and not more than 44 percent water. But the cheese maker can substitute skim milk, buttermilk, and whey for cream and butterfat. Otherwise it's pretty much the same as American cheese.

Cheese spread, which is what Velveeta now calls itself, can contain up to 60 percent water and not less than 20 percent butterfat. It may also contain small amounts of gums for body, and sweeteners.

The name in processed cheese is J. L. Kraft, founder of the modern food conglomerate Kraft General Foods. Kraft began his cheese career in 1900 at Ferguson's general store in Fort Erie, Ontario. He was the cheese clerk, selling to housewives from a sixty-pound Cheddar that was kept behind the counter under a bell glass. His cheese clerking led to a job with a cheese company in Buffalo. On business in Chicago he saw an opportunity. Merchants there were making daily trips to the cheese market. Kraft took sixty-five dollars in capital, leased a horse named Paddy, and began a cheese route, saving store owners the trouble of going to the cheese market. His route was so successful that in 1909 he and his two brothers formed J. L. Kraft Brothers Company to manufacture cheese, in addition to selling it.

His early experiments involved heating cheese and forming cheese blends. He developed the first commercially feasible method of processing cheese and packaging it in hermetically sealed, completely sterilized containers.

In 1916 he was awarded a patent for the "process of sterilizing cheese and an improved product produced by such process," which effectively turned cheese into a uniform dough.

The Kraft advance that made Velveeta possible came in 1921, when Kraft patented a method of packing processed cheese in tinfoil-lined wooden boxes, a great improvement over canned cheese, which was popular at the time. Foil had been used for packing cheese since 1904, but Kraft's genius was in developing a foil that would stick to the cheese and not the box, creating an hermetic seal. He brought out a five-pound loaf in 1921 that was a sensation. Consumers liked it because it didn't get stale easily. And grocers liked it because it was easier to handle.

1 DOZEN GRADE A LARGE SUPERBRAND EGGS—$0.78

If it weren't for the egg, we wouldn't have omelets or egg salad or egg fu yung. Or sex. That's right. We can thank the egg for sex. Or, more precisely, the eggshell.

In prehistoric days, when the only life forms were fish and amphibians, eggs were laid in water and the sperm fertilized them there. Then in the evolutionary transition from amphibians to reptiles, some lower form of life had to invent sex. Sex was necessary because the eggs had to be fertilized before the shell could form: That means internal fertilization. Sex.

Dr. Alfred S. Romer, a paleontologist at Harvard's Museum of Comparative Zoology, calls the egg "the most marvelous single invention in the whole history of vertebrate life." And that includes Oprah. The egg is a perfect enclosed life-support system with everything the embryo needs for growth and maintenance until hatching.

The egg has always been an object of curiosity, and not just the old which-came-first-the-chicken-or-the-egg curiosity either.

For centuries men and women wondered how an embryo could flourish with no outside support and no air to breathe. It was left to John Davy of Edinburgh, Scotland, to demonstrate in 1863 that an egg could breathe. He placed an egg underwater, then filled it with pressurized air and watched as bubbles popped out over its entire surface. The shell wasn't solid as everyone believed but had pores to allow carbon dioxide to pass out and oxygen to pass in. It wasn't exactly the discovery of America, or even the discovery of NutraSweet, but it sure beat arguing about which came first, the chicken or the egg.

In 1990 there were more than 250 million chickens in the United States, which means there are more of them than there are of

---

FORKLORE #34

KIDS, DON'T TRY THIS AT HOME

You know the expression "walking on eggshells," but did you know it's only an expression, that eggs aren't nearly as fragile as people assume?

Consider the egg-dropping craze that swept England in 1970. A schoolmaster started it by demonstrating the egg's resilience to his students. He dropped a dozen out a second-story window and none broke. Then a fireman at a nearby fire hall dropped ten eggs from the top of a 70-foot ladder. Seven survived. Next a Royal Air Force officer released a dozen and a half from a helicopter hovering 150 feet aboveground. Only three broke.

It was left to a newspaper to perform the ultimate experiment. The *Daily Express* hired a pilot to dive-bomb a field at 150 miles an hour, releasing five dozen eggs as he buzzed the ground. Only three broke.

Still unconvinced? Try this experiment. Put an egg endways between your palms and push. Hard. Harder. See, you didn't even need to do it over the sink. Maybe Hulk Hogan could crush it, but the average person can't.

The astonishing strength of the egg comes not so much from its shell thickness as from its design: two convex arches. As you push, the force is distributed evenly downward and outward.

---

us. That is, if you call them chickens. Egg machines may be more like it. It is a rare chicken that scratches around the henhouse, retiring at night to lay an egg. It is a rare chicken that scratches anything except perhaps one of the other three chickens in its tiny cage. It is a rare chicken that even sees daylight. Most hens are raised in controlled-environment chicken houses, where their sole job is to eat and lay eggs.

And they are good at it, far better at laying eggs than their chicken ancestors. In 1800 the average barnyard hen laid about 15 eggs a year. Selective breeding has increased that to about 300 eggs a year. And a leghorn named No. 2988 at the University of Missouri College of Agriculture laid 371 eggs in 1980. That's better than an egg a day. Whew!

Selective breeding has also produced a more efficient chicken. In the early fifties a six-pound hen ate about nine pounds of feed for every dozen eggs she laid. Now that is down to four pounds of feed.

And breeders are working on improving on all of that. They are always on the lookout for a chicken that can reach sexual maturity a little earlier, that can weigh a little less and eat a little less feed and still produce the same amount of eggs, maybe even more eggs.

It's economics, not demand, that is fueling this selective breeding. Egg consumption has declined steadily since 1944. In 1967 Americans ate 332 eggs per person. By 1990 that had fallen to 234 eggs per person. And that included the eggs in packaged foods such as Hostess Twinkies and chocolate chip cookies. Eggs for breakfast is a thing of the past. Working moms dump cereal in a bowl instead.

We're buying Grade A large eggs. The grade tells how fresh the eggs are. Grade A means they were packed in the last thirty days. Grade AA eggs were packed in the last ten days. There are no Grade AA eggs in the store.

There are six sizes of eggs: jumbo, extra large, large, medium, small, and peewee. You seldom see anything below medium in the supermarket. Large means a dozen weigh twenty-four ounces.

These are brown eggs. Brown eggs don't come from brown chickens. Necessarily. They can. Brown eggs just have a pigment they picked up in the chicken's oviduct. They don't have thicker shells or better yolks. They are exactly the same on the inside.

16 OZ. TUB LAND O LAKES SPREAD WITH SWEET CREAM—$1.24

This "spread" we're buying today isn't butter. In fact, it doesn't even meet the federal standard of identity for margarine, so it has to be called spread. One look at the ingredients and you know this is not Grandma's butter: liquid soybean oil, water, partially hydrogenated soybean oil, cream, skim milk, salt, vegetable mono- and diglycerides, soy lecithin, potassium sorbate as a preservative, lactic acid, artificial flavor, vitamin A palmitate, colored with beta carotene.

If I didn't already know the story, I would be greatly concerned with that last ingredient: "colored with beta carotene." They colored my butter? In fact, they've been coloring butter and margarine for more than a hundred years. Butter was the first food approved by Congress to have added color. That was in 1886.

Margarine—which is this "spread's" closest relative—is naturally white: It looks like lard. If it weren't that airy yellow color, I might have trouble spreading it on my toast. Yuck.

I've grown accustomed to my margarine being yellow.

But when oleomargarine was introduced in this country in 1873, it was illegal for manufacturers to add any coloring that would lead shoppers to confuse it with butter. That was the butter makers speaking through their elected representatives. So oleo makers included a packet of yellow coloring in each package of margarine and the majority of buyers would mix it in. They didn't want a white lardy-looking spread.

The coloring used in oleo a hundred years ago is the same kind used in my Land O Lakes spread today, a carrot-based yellow. Most all butters, margarines, and cheeses are colored with some sort of carotene color, extracted from carrots.

Beta carotene is what's called a natural color.

The government recognizes three kinds of food colors: those naturally present in foods, those present in nature that may be added to foods (like beta carotene), and artificial or "coal tar" dyes, so called because the earliest ones were made from coal by-products. Modern artificial colors are no longer extracted from coal products; they are made from chemicals, right out of a chemistry set.

Artificial and natural colors are everywhere: Just look. The tomato in frozen pizza is colored with beet powder. Soft drinks are colored with caramel. The cherries in fruit cocktail are bleached with sulfur dioxide, then colored with red dye.

And the colors are there for one reason: Shoppers want food that is attractive to look at.

In 1972 food scientists decided to study the importance of food color. They knew it was important, but how important? A group of twenty subjects were served a plate of steak, French fries, and peas. They devoured the banquet when eating under special lighting that concealed the fact that the food had been colored. When the room was returned to normal lighting, revealing blue steak, green French fries, and red peas, a number of the subjects reported feeling ill and were unable to continue eating. Since then the food industry has spent billions of dollars making food look better.

1 LB. BLUE BONNET MARGARINE—$0.53

Oleomargarine was invented in 1809 by a chemist's assistant who thought he was doing something else. Hippolyte Mege Mouries had earlier developed the effervescent pill. His invention of margarine was based on a misconception: that the cream in milk comes from beef suet. He heated suet to 30°–40° C. and with pressure was able to extract a small bit of suet that would melt between 20° and 25° C. The first margarines were made exclusively from beef tallow. As margarine production increased there wasn't enough tallow to keep up. By 1874 some manufacturers were substituting a small amount of vegetable oil. In 1907 it was discovered that solid vegetable fats also worked.

In 1869 the French scientists P. Sabatier and J. B. Senderens figured out the process of hydrogenation: hardening vegetable oils. A German chemist, G. Norman, patented the process. Since hydrogenation worked on virtually any oil, margarine manufacturers were no longer tied to beef tallow. By 1911 almost none used it.

Over the years margarine has gradually replaced butter on the table. Butter outsold margarine almost two to one in 1949. By 1989 that ratio had reversed.

Judy has now covered every aisle during this trip except Health and Beauty Aids, Housewares, and Baby Products. She has covered about 75 percent of the store. According to a 1991 Food Marketing Institute survey, a shopper normally will cover only 41 percent of the supermarket on an average trip.

We're both tired; we're ready to head for the checkout. But we're not done yet!

The Back Aisle:
Meat

▲■▲■▲■▲■▲■▲■▲■▲■▲■▲■▲■▲■▲■▲■▲■▲■

The modern consumer protection movement was born in the meat department, amidst ground chuck and California hams and head cheese. And it was all an accident. Upton Sinclair, a twenty-six-year-old journalist, was sent to Chicago in 1904 by the Socialist newspaper *Appeal to Reason* to investigate the city's giant meat-packing industry. He found the stockyards and meat-packing plants overrun by filth and squalor, and the packers scheming to cut every corner to satisfy their greed. But it wasn't the outrage of adulterated meat and filthy packing rooms or the crooked meat packers that upset Sinclair. It was the oppression of the workingmen and -women employed by the plants. And that is what he thought his novel, *The Jungle*, was about. But, as he would later say, "I aimed at the public's heart and by accident I hit it in the stomach."

It was passages like this one that hit the public's stomach:

With one member trimming beef in a cannery, and another working in a sausage factory, the family had a first-hand knowledge of the great majority of Packingtown swindles. For it was the custom, they found, whenever meat was so spoiled it could not be used for anything else, either to can it or else to chop it up into sausage. . . . Jonas had told them how the meat that was taken out of pickle would often be found sour, and how they would rub it up with soda to take out the smell, and sell it to be eaten on free-lunch counters; also of all the miracles of chemistry which they performed, giving to any sort of meat, fresh or salted, whole or chopped, any color and any flavor and any odor they chose. In pickling hams they had an ingenious apparatus by which they saved time and increased the capacity of the plant—a machine consisting of a hollow needle attached to a pump; by plunging this needle into the meat and working with his foot, a man could fill a ham with pickle in a few seconds.

And yet, in spite of this, there would be hams found spoiled, some of them with an odor so bad that a man could hardly bear to be in the room with them. To pump into these the packers had a second and much stronger pickle which destroyed the odor—a process known to the workers as "giving them thirty percent." Also, after the hams had been smoked, there would be some that had gone to the bad. Formerly these had been sold as "Number Three Grade," but later on some ingenious person had hit upon a new device, and now they would extract the bone, about which the bad part generally lay, and insert in the hole a white-hot iron. After this invention there was no longer Number One, Two, and Three Grade—there was only Number One Grade.

The Jungle was serialized, to little note, in *Appeal to Reason* in 1905. The next year it was published as a novel and public outrage could be heard from Washington State to Washington, D.C. And before the year was out Congress had rushed through the first national food law, the Pure Food and Drug Act of 1906.

The consumer movement was born and its accidental father was Upton Sinclair, a religious Millerite given to periods of fasting for "self-purification."

The Pure Food and Drug Act of 1906 marked the first federal law in this country to address the food supply system. But food legislation dates back much farther.

One early instance of a food manufacturer's trying to sneak one past the public was in 1202 and it prompted King John of England to sign a decree, the Assize of Bread. It seems a few of the bakers at the time were trying to bulk up their bread with cheaper ingredients, such as peas and beans. King John's decree made that practice illegal, setting the penalty at flogging. If they enacted a law in 1202, then you can be sure the practice dates back long before that.

The first food law in this country was a Massachusetts statute, passed in 1784, that prohibited the sale of unwholesome provisions. The Illinois legislature passed a law in 1874 prohibiting the addition of "inferior material" to food or mixing diseased substances with sound and concealing them by packing. As early as 1879 a pure food bill had been introduced in Congress, but it died in the Committee on Manufactures. New York and New Jersey both passed pure food laws in 1881, but the laws weren't strictly enforced.

All this legislative rumbling was a reaction to a very real problem. In the nineteenth century the food industry was a rough-and-tumble business and when it came to purchasing food, the rule was caveat emptor. Among the deceptions commonly practiced at the time were plaster of Paris sprinkled into flour, starch mixed into cocoa, and cottonseed meal stirred into mustard. Ground pepper was adulterated with charcoal and what was called "pepper dust," which was nothing more than the sweepings of the storeroom floor. Cheeses were colored with lead and arsenic. Raspberry jam might be a raspberry jam base with glucose, hayseed, and artificial colors and flavors added. It was then, as it is today, a matter of economics. Ground charcoal was cheaper than pepper. A pound of pepper could be adulterated with charcoal and floor sweepings and sold as a pound and a half of pepper.

Advances in chemistry had produced a number of healing chemicals including germicides and antiseptics. Food processors figured if these chemicals saved living tissue, they could preserve dead tissue. The result was the wholesale dumping of medicinal chemicals into food products.

The war against deceptive practices in the food processing industry heated up in 1883 with the elevation of Dr. Harvey W. Wiley to chief of the U.S. Department of Agriculture's Bureau of Chemistry, the forerunner of the Food and Drug Administration. Sinclair may have been the accidental father of food laws, but it was Wiley who paved the way. Upon ascending to his job, Wiley began a study of the presence of adulterants in food. He was a ferocious defender of the food supply but lacked the tools to attack the problem. He had to challenge each food adulterator in court, case by case. He won some; he lost some.

He tried to alert the public to the problem of adulterants with the creation of what he called Poison Squads, panels of healthy, young male volunteers. His squads ate only from a controlled list of foods. From time to time preservatives or other adulterants were added to their diet as an experiment. One Poison Squad report noted, "The administration of benzoic acid, either as such or in the form of benzoate of soda, is highly objectionable and produces a very serious disturbance of the metabolic functions, attended with injury to digestion and health." But his reports produced no large-scale public outcry. Not even *Bulletin* 25 of the Division of Chemistry, published in 1890, which warned that virtually every food on the market was adulterated.

Adulteration needed to be dramatized to get the public's— and Congress's—attention. Upton Sinclair did that. And Congress reponded with the passage of the Pure Food and Drug Act on June 30, 1906:

"An Act for preventing the manufacture, sale or transportation of adulterated or misbranded or poisonous or deleterious foods, drugs, medicines and liquors and for regulating traffic therein, and for other purposes."

The law was well intentioned, aimed not only at food adulteration but also at deceitful labeling.

"It shall be unlawful for any person to manufacture within any Territory or the District of Columbia any article of food or drug which is adulterated or misbranded within the meaning of

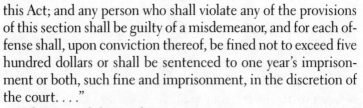

this Act; and any person who shall violate any of the provisions of this section shall be guilty of a misdemeanor, and for each offense shall, upon conviction thereof, be fined not to exceed five hundred dollars or shall be sentenced to one year's imprisonment or both, such fine and imprisonment, in the discretion of the court. . . ."

What was adulterated?

Well, . . . "in the case of confectionery, if it contains terra alba, barytes, tale . . . ," whatever those things are (and that last one really worries me).

And . . . "in the case of food . . . if it contains any added poisonous or other added deleterious ingredient which may render such article injurious to health [or] if it consists in whole or in part of a filthy, decomposed or putrid animal or vegetable substance or any portion of an animal unfit for food, . . . or if it is the product of a diseased animal or one that has died otherwise than by slaughter."

The law was also specific on what was deceitful labeling. The most worrisome was this section:

"In the case of food . . . if it fails to bear a statement on the label of the quantity or proportion of any morphine, opium, cocaine, heroin, alpha or beta eucane, chloroform, cannabis indica, choloral hydtrate or acetanilide or any derivative or preparation of any of such substances contained therein."

It was a start. But an act conceived in haste in 1906 could not possibly foresee all the additives and deceptions that the food industry could invent in the coming years. By the mid-twenties the 1906 law was outdated and outpaced. It was full of ambiguous language that didn't allow the government to set standards for food composition or set levels of excessive contamination. Wiley, who had retired from the Bureau of Chemistry in 1912 to write a column for *Good Housekeeping*, noted in his 1929 book, *History of a Crime Against the Food Law*, that food processors had learned how to evade the law. The bureau was partly to blame for the problem. After Wiley retired, enforcement of the act shifted from punishment to education.

Congress began working on a new food law in 1933. The efforts were spurred—again—by a book. This one was called *100,000,000 Guinea Pigs* and it was written by a couple of engineers, Arthur Kallet and F. J. Schlink. Their book, a slashing diatribe, was more concerned with reform than with reality. Kallett and Schlink's thesis was that Americans were "all guinea pigs and any scoundrel who takes into his head to enter the drug or food business can experiment on us." Before Congress could stall on any new food reform, it was hit with a second exposé, *American Chamber of Horrors*, published in 1936. And this time the author couldn't be waved off as a kook. The book was written by an FDA employee, information officer Ruth Lamb, who made extensive use of the bureau's files. Her longest chapter, on the problems of pesticide spray residue on fruit, detailed the deaths of a number of children from pesticide poisoning. "The real risk from arsenic and lead in spray residues is not acute poisoning! Rather the danger lies in the slow, insidious undermining of health from the accumulation of metals in the soft tissues and bones."

So on June 25, 1938, Congress passed the Food, Drug and Cosmetic Act, which gave increased regulatory authority to the Food and Drug Administration:

"Section 301: The following acts and the causing thereof are hereby prohibited:

"(a) The introduction or delivery for introduction into interstate commerce of any food, drug, device or cosmetic that is adulterated or misbranded.

"(b) The adulteration or misbranding of any food, drug, device or cosmetic in interstate commerce."

The cornerstone of the law was that it provided for the Secretary of Agriculture to establish legal definitions for common processed foods. With such standards of identity it would be much easier to determine if a food processor was adulterating a product.

The 1938 law was a quantum leap in food regulation: It gave the Secretary of Agriculture broad power to control shipments of sprayed fruits and it required manufacturers to label foods containing artificial colors and preservatives. The key provision allowed the Secretary of Agriculture to set up standards of iden-

tity for common foods to alleviate a common problem: manufacturers who scrimped on ingredients.

But the law had its flaws. For one, it was incumbent upon the
FDA to prove an additive was unsafe. Twenty years later, in 1958,
the overburdened FDA petitioned Congress to reverse the procedure for approving a food additive. The 1958 Food Additives
Amendment to the 1938 law made it the responsibility of the
manufacturer to show that an additive was safe. The 1958 amendment also recognized three categories of food additives: those
previously sanctioned as safe, those generally recognized as safe
(GRAS), which included some 700 substances including caffeine, cinnamon, MSG, mustard, and sugar, and all other additives. Only the latter category of foods was subject to regulatory
sanction.

The most recent food law was the 1977 Delaney Clause, an
amendment to the 1958 amendment to the 1938 law. The food
industry doesn't like it because it decrees that no additive is permitted if it has been shown to cause cancer in humans or ani-
mals.

How could you be against a law like that?

HOLLY FARMS CHICKEN BREASTS—$3.99

The chicken in the supermarket may say it comes from a bu-
colic-sounding place like Holly Farms, but the modern chicken
processing plant isn't a chicken farm so much as a chicken factory. Pork farmers have an expression for their all-in-one farms
and packing plants: "From the squeal to the meal." Chicken
farmers might use this one, if it weren't so gross: "From the yolk
to the yuck."

The chicken, not the egg, comes first on the commercial
chicken farm assembly line. Rather, thousands of chickens. One
breeder house will hold 9,000 hens and 900 roosters just for the
purpose of breeding. And a good-sized farm may have as many
as thirty-two houses producing a quarter million eggs a day.

These aren't just any chickens laying these eggs either. They're superbirds, what the industry jokingly calls Dolly Parton birds. They have larger breasts than the chickens of old because today more people want white meat than dark. The average hen of today weighs 5.15 pounds and takes 46 days to reach maturity. In 1968 the typical hen weighed 3.5 pounds, took 102 days to reach market size, and ate 15 percent more feed doing it.

Most of these breeding hens—85 percent—will lay an egg a day. Workers collect the eggs daily and send them immediately to the hatchery next door. There they are shoved into the incubator room for eighteen days. At the end of that period they are moved—135 eggs to a basket—to the hatching room. In three days the eggs begin hatching. There are chicks hatching every second. It's like Easter, chicks everywhere. It's almost impossible for workers to keep up. They scoop up handfuls of chicks and toss them onto conveyor belts to be taken to the inoculation station. There the birds are immunized against disease. In some farms they are also debeaked—their beaks are blunted on a hot plate to reduce the danger of hurting other chicks.

FORKLORE #35

NEVER ASK A CHICKEN FOR THE TIME

Chickens may live among us and share our dinner table, but they do not share our biological clock. A chicken day is twenty-six to twenty-eight hours. Researchers have tried over the years to use this anomaly to produce more and bigger eggs. A researcher at Cornell exposed chickens to two hours of light, six hours of dark, two more hours of light, then eighteen hours of dark. The result was increased egg production and bigger and thicker-shelled eggs. The eggs were heavier by an average of six grams and stronger by 10 to 12 percent.

Egg men will vary periods of light and dark in their henhouses for various reasons: to increase production, to increase egg size, to increase thickness of the shell, to standardize the hours for egg collection, even to level out the light bills. But probably never because the chickens want to sleep.

From the "throwing room"—so called because the baby chicks travel several feet in the air when they are thrown onto the conveyor belt—it's on to the grow-out house, a 40-by-400-foot barn where 23,000 chickens are fattened at a time.

At the end of six weeks the birds are caught and shipped—twenty-two to a cage—next door to the slaughterhouse, where the cages are dumped onto yet another conveyor belt. Workers grab the birds and hang them upside down by hooking their feet in a moving shackle. An experienced worker can hang a bird a second. The birds are sprayed with water, then the shackle guides them past an electrically charged metal grate. The narrow path makes it impossible for the bird not to brush up against the eighteen-volt charge. The shock isn't enough to kill the chicken, just enough to stun it. As the stunned chicken hangs limply, a mechanical blade severs its throat. Blood sprews out and into a floor drain.

The chickens continue along the conveyor, blood dripping down. By the time all the blood has drained out, the conveyor reaches the scalding pool and the dead chickens are dragged through 135°F. water to loosen their feathers. The feather-removing machine—with six-inch spinning rubber fingers—flogs away at the birds, knocking the feathers onto the floor, where they are washed down the drain.

The worst job at a chicken processing plant is cleaning the drains.

The dead birds continue down the line where a worker or a machine cuts off the heads, slices open the body cavities, and removes the entrails. It is here that U.S. Department of Agriculture inspectors check the birds for tumors, infections, and other diseases. An inspector has about two seconds per bird. Diseased birds are pulled and destroyed. If a bird has only one tumor, the spot can be cut off and the bird salvaged. Two or more and the bird is condemned. Those that pass are trimmed, washed, and chilled in a tank with 5,000 other dead chickens.

From here the birds head to the processing plant. There the chickens are weighed and graded by size. Workers standing shoulder to shoulder cut up the birds headed for specialty packages.

Others are sold as whole chickens. The packages are wrapped and shipped out to supermarkets in refrigerated trucks.

The entire poultry industry shipped half a billion pounds of processed chicken each week in 1990. Chicken consumption was twenty-four broilers per person for the year.

If your chicken is yellow, a coloring agent was added. It makes the pale-skinned carcass look more attractive.

4.8 OZ. PLUMP'N TENDER TURKEY BREASTS—$7.18

In the supermarket you encounter many misnamed products: Froot Loops contain no fruit. Mrs. Smith's Natural Juice Apple Pie is full of artificial ingredients. Chee-tos are made from corn-meal.

So it's nice to know that at least one product lives up to its name: The turkey is a turkey.

Don't let anyone kid you; the turkey is not a genius species. The turkey is dumb. John Hudson, a Virginia turkey farmer, told me turkeys have been known to starve to death while surrounded by mountains of food simply because it didn't occur to them to eat. He said his family moved its turkeys indoors because they had a bird stare up at the rain and drown.

Even the turkey's rise to most-favored status on the Thanksgiving menu came in a roundabout way. When the Spaniards invaded Mexico, they found the turkey being bred in captivity. They exported the bird to Spain and from there to England and France. When the Pilgrims came to America, they brought the bird back with them. The turkey was already a staple for English Christmas feasts, so it was only natural that the Pilgrims would use the turkey for the first Thanksgiving.

If Benjamin Franklin had had his way, the turkey, not the eagle, would be the national bird. Franklin reasoned that the turkey, by virtue of the fact that it was native to every state in the union (thirteen at the time), should represent the country on the national seal. Fortunately wise heads prevailed. Can you imagine

Boy Scouts striving to become Turkey Scouts? Or the 101st Airborne Division flying into battle as the Screaming Turkeys?

Rockingham County, Virginia, modestly refers to itself as the turkey capital of the world. More turkeys are grown and processed there than any other place in the world. Each year the Poultry Association sponsors the Friends of Feathers Festival in Harrisonburg. They hold a parade with turkey-related floats and crown the poultry queen, a title any Rockingham County girl would be proud of.

Rockingham County is the home of a number of major turkey farms, virtual perpetual-motion machines. Here turkeys are raised without ever seeing the light of day or touching the earth. Turkeys are fattened in wire cubicles in giant barns. When fully grown, they are hung on a conveyor and slaughtered, scaled, plucked, processed, and frozen.

The turkey's digestive system is a marvel. Grand Duke Ferdinand II of Tuscany, with apparently nothing much to do, forced turkeys to eat glass balls, hollow lead cubes, and wooden pyramids, to test the bird's gizzard. The next day, when the experiment was concluded, the turkeys were found to have crushed the glass to a powder, flattened the lead cubes, and worn down the wooden pyramids.

Columbus was the first European to see a turkey—and also the first to eat one. He was hospitably received by the Honduran natives on August 14, 1493, and treated to a feast of native fowl including the turkey.

No one knows how the turkey got its name. The name turkey was in use in Europe long before the present bird was found. At that time it referred to a peafowl. In the sixteenth century the turkey was often confused with the guinea fowl and both were referred to as turkeys. Some have suggested that the bird was so named because it was thought to come from Turkey. Others think the name comes from the bird's call note, which they say sounds like "turk, turk, turk." Other derivations range from the Hebrew word for peafowl, *tauas*, to the Malabar word, *togei*. The Indians had more than twenty names for the turkey, none of which was turkey.

Eldridge Hawks, who works at City Poultry in my hometown, told me the story of the time he was unloading a truckload of frozen processed turkeys at a Rogersville, Tennessee, market when a flock of turkeys gathered around his truck—to watch.

I LB. GROUND CHUCK—$2.09

For years our Winn-Dixie advertised itself as "the beef people." They even had one of those singing commercials where a chorus of tenors entoned, in harmony, "Wiiiinn Dixie . . . the beeeeef people," as if that rhymed or something.

But red meat consumption is declining. Cholesterol is part of it. Economics is another part; chicken is cheaper. And now Winn-Dixie just calls itself "America's supermarket."

The beef industry has tried valiantly to reverse the trend. It even hired Cybill Shepherd and Jim Garner to tout beef in a series of TV commercials. That sort of backfired when ol' Jim had to go in the hospital and get his arteries cleaned out. Too much beef, everybody said.

I can't remember the last time we had a roast.

I LB. PACKAGE OSCAR MAYER
HOT DOGS—$1.99

The hog dog is a sausage and sausage dates from 900 B.C. Sausage (but not the hot dog specifically) was even mentioned in Homer's *Odyssey*.

During their history hot dogs have been called frankfurters, for one of their towns of origin, Frankfurt, Germany, and wieners, for another of their towns of origin, Vienna, Austria. In this country they were called dachshund sausages, because of their resemblance to the dog.

They were first sold here on Coney Island in the 1880s by a German immigrant named Charles Feltman. Feltman started out hawking pies on the beach, but when the hotels opened their own restaurants, his pie business fell off dramatically. So he switched to the frankfurter. From a simple hot dog stand he built a food empire worth $1 million at the time of his death in 1910.

The name hot dog originated on a cold April day in 1901 at New York's Polo grounds. The New York Giants were playing a baseball game, but the weather was so nippy that no one was buying ice cream or cold drinks. Concessionaire Harry Stevens sent his hawkers out with dachshund sausages, instructing them to yell, "Get 'em while they're hot." Newspaper cartoonist Tad Dorgan was in attendance and drew a cartoon about Stevens's solution to the cold front. But Dorgan didn't know how to spell *dachshund*, so he called it a hot dog.

Hot dogs in an elongated bun were a sensation at the 1904 St. Louis World's Fair, the same fair where peanut butter and ice cream cones made their debut. That's one fair I wish I had been at.

Despite its Coney Island origins, the hot dog was banned from the island in 1913 by the Chamber of Commerce, which was reacting to rumors that the wiener—true to its name—was made from ground canine. (See, those rumors didn't start with this generation.)

The standard size for one hot dog is 1.6 ounces, so when they were introduced into supermarkets in the thirties, a ten-pack conveniently weighed in at 1 pound. Buns have traditionally been sold in eight-packs and twelve-packs, creating the eternal problem of juggling the number of packs of hot dogs with the number of packs of buns. You had to have a math degree from MIT just to figure out how to make your picnic come out even.

Judy and I don't have that problem today. This new "bun-length" beef frank is a bit larger: eight to a pound. Meaning it matches the buns perfectly: in length and in packaging.

It was Otto von Bismarck who supposedly said the two things you don't want to see made are sausage and law. I've seen C-SPAN, cable TV's Congressional Coverage, and that was enough for me. I never plan to visit a hot dog factory.

The hot dog of today is only a distant relative of Charles Feltman's frankfurters. Feltman made his from chunk beef, ground lean pork, pork juice, pepper, salt, sugar, ginger, nutmeg, paprika, and the Ceylonese spice *corcandes*.

Here's what Oscar Mayer puts in ours: beef, water, salt, corn syrup, dextrose, flavoring, autolyzed yeast, sodium erythorbate, oleoresin, paprika, sodium nitrite.

The beef in our hot dog is not USDA Grade A select prime rib. It's the scraps and trimmings of other cuts. But it wasn't steak in Feltman's day either.

Water replaces pork juice to keep the hot dog juicy. Water is a lot cheaper. The average hot dog is one-fourth water. Hey, they have salt in common. And don't forget the sugar, although Oscar Mayer is using cheaper corn syrup and dextrose.

While Charles Feltman's hot dogs were seasoned with natural spices, Oscar Mayer's hot dogs derive some of their taste from "flavoring."

Sodium erythorbate is a preservative to give the hot dog a longer shelf life. Sodium nitrite is a triple-threat additive. It inhibits the growth of the spore that produces botulism. It gives the meat a tangy taste. And—most important to processors—it reacts with the myoglobin molecule to give the hot dog a meaty red color.

Feltman probably didn't need preservatives; he sold his hot dogs as fast as he could make them.

Oleoresin paprika is what the Food and Drug Administration calls a natural color. It's what gives our otherwise gray meat its red "hot dog" color.

7 OZ. SPAM—$1.15

Spam has been the butt of jokes for half a century. During World War II, GI's joked that Spam was just ham that couldn't pass its Army physical. During the eighties David Letterman spoofed it with the new Spam-On-A-Rope—"in case you get hungry in the shower."

Spam knows it's funny. Less-salt Spam was introduced with a mock-serious radio commercial that compared it to New Coke and promised that if the new version of Spam caused the same furor as reformulated Coca-Cola, "we still have classic Spam."

Spam is funny. Even the name's funny: Spam.

Almost ham. And that's really where the name came from. It's spiced ham. The name was created in 1937 at Geo. A. Hormel and Company for their new line of canned lunch meats. The company sponsored a contest to pick the name, with the winner getting $100. Among the suggestions were Brunch and Hamlet. The winning entry belonged to Donald Douglas, the actor-brother of Hormel vice-president James Douglas.

Spam gained popularity during World War II, when it was a staple in the rations passed out to U.S. troops. Many soldiers, my father included, came home with a taste for the stuff. Spam is no longer included in military rations, but it did go back to war in 1991. The Pentagon sent about $2 million worth of Spam into the Gulf War.

What is Spam? Spiced ham. Really. It's scraps of chopped pork shoulder and chopped pork ham—with lots of fat; witness 79 percent of its calories come from fat—with salt, water, sugar, and the preservative sodium nitrite added.

But Spam's popularity is nothing to joke about: Americans eat 113 million cans a year. It's particularly loved in Hawaii (12 cans per person per year) and Alaska (6 cans per person per year). The next closest states are Texas, Alabama, and Arkansas with 3-can per capita averages.

Spam's manufacturer, Geo. A. Hormel and Company, has grown past its origins as a meat packer. It was founded by George Hormel in 1887. Four years later the company introduced its first new product: cured pork loin with the backbone removed. Today Hormel calls that Canadian bacon. Hormel was strictly a meat packer until 1979. At that time about 70 percent of its sales came from raw-meat products. But in the decade plus since then Hormel has rolled out scores of new products, from Top Shelf microwave dinner entrées to Health Selection microwaveable meals. Today about 70 percent of Hormel's sales come from brand-name food products instead of raw meat.

And what was the name of Hormel's first brand-name consumer product? Spam, of course.

The Checkout Lane

Here we are at the checkout line and it's only 7:18 P.M. Our shopping excursion took forty-eight minutes. That's a little quicker than the average.

In the time we've been here the store has probably done $1,690 in business. That's what the average supermarket does in an hour.

There are two other shoppers ahead of us. As we wait Judy browses through the magazines. She ends up tossing the latest *Women's Day* in the cart. Then, when she sees the batteries above the candy, she remembers our son's Game Boy needs a new battery, so she adds a pack of batteries to our order. Which is exactly what the supermarket's designers had in mind. The checkout lane is the home of impulse items: cigarettes and candy bars, magazines, nail files and batteries, the kinds of things no one would make a special trip to pick up. But they're cheap enough that many of them end up in the cart.

WEEKLY WORLD NEWS—$0.50

If Judy can add a women's magazine, then I can add one of my favorites. I toss in the latest issue of the *Weekly World News*, a screaming tabloid that was created as a by-product of the *National Enquirer*. That doesn't mean that it gets all the stuff that is too scandalous for the *Enquirer*. There is no such thing. No, when the *Enquirer* went to full color, it still had that old black-and-white printing press. What to do? Create another supermarket tabloid.

I buy the *Weekly World News* whenever they have a great headline, like "World War II Bomber Found on Moon!" or "Man's Breath Kills Wife!" After all, these are the guys who spawned the "Elvis is alive!" controversy, the story that, unlike its subject, refuses to die.

I once read an interview with one of the editors in which he spoke candidly, perhaps too candidly, about how they get their stories. They don't make them up! But if someone calls in and says, "I just saw Elvis at the Burger King!" they don't ask, "How do you know it was Elvis?" They ask, "Which Burger King?"

The headline today is a classic: "UFO Lands on Navy Carrier!" How did the local paper miss that story?

BAZOOKA BUBBLE GUM—$0.05

When I was a kid, there was Bazooka Bubble Gum and Topps baseball card bubble gum and that was it. And they were both made by Topps, so the only difference was in shape; Bazooka was in a wad about the size of the end of your finger, and baseball card bubble gum was in a stiff flat stick and it always seemed to stick to the bottom baseball card.

Now bubble gum comes in every imaginable size, shape, and taste. There's Big League Chew (shredded gum sold in a pouch like chewing tobacco) and Tape Chewing Gum (a six-foot roll of flat bubble gum in a small tin) and Popeye Gum (looks like spinach but, fortunately for kids and Popeye Gum sales, tastes like bubble gum) and Tubble Gum (bubble gum in a squeeze tube).

Most bubble gum is sold to children between the ages of six and seventeen. All the varieties of gum available today have increased consumption somewhat, but basically the bubble gum market depends on the birth rate. The more kids there are, the more bubble gum is sold.

The best-selling bubble gum flavors are fruit, grape, strawberry, watermelon, and cola. Flavors that have bombed in the bubble gum market include banana, chocolate mint, and tangerine.

Bubble gum makers keep introducing weird-shaped bubble gum: Amurol Products Company, a subsidiary of Wrigley and the makers of Hubba-Bubba, even tried an Ouch gum in the shape of a Band-Aid. What a good idea; how could that have failed?

Kids—bless their hearts—keep coming back to the old chunk, the basic Bazooka shape. That's because the chunk shape produces the biggest bubbles.

More gum is sold in the Southeast and most experts attribute that to the sugar in the gum. Food items with lots of sugar tend to sell better in the Southeast than in other parts of the United States.

At 7:26 we are at the cash register. That's eight minutes in line, again about average for a major shopping trip.

The cashier smiles a quick smile, adds "How are you this evening?" then begins speeding each item across the laser scanner, waiting for the electronic blip sound before running the next item across. At some stores the cashier can earn a bonus if he or she averages a certain number of items per minute. Other stores let shoppers vote on the employee of the month and then give the winner a cash award and a special parking place. Our oldest son, who used to work at Winn-Dixie, says they don't have any incentive programs.

• • •

I am watching the items blaze across the register display, when it comes, from the bag boy.

In the nineties the most dreaded nonpersonal communication has shifted from "Hi, my name is Roy, I'll be your bagger" to "Paper or plastic?"

After an hour of making one decision after another, you'd like to rest it at the checkout lane. But no, we have to decide whether to have our groceries packed in good old-fashioned paper bags—kraft bags, they are called in the trade—or newfangled thin plastic bags.

We try to weigh the environmental consequences—plastic is better, we are told, because it can be recycled—against the practical ones: We use paper grocery bags around the house for a number of chores.

It is now this way in nine out of ten supermarkets; you have to choose between paper and plastic. Ninety-nine percent of stores have paper bags; 88 percent have plastic.

It has only been in the last two years that Winn-Dixie has offered plastic bags. Not that the plastics industry hasn't been trying to enter the lucrative grocery-bag industry for years. There are 22 billion grocery sacks used a year.

The Celloplast company introduced the first plastic grocery bag in 1972, but neither shoppers nor clerks liked it. Maybe it was strong enough to handle all those canned goods, but it sure didn't look like it. That bag was a bust in supermarkets, but it is still in use today. It's called a T-shirt bag; it was readily adopted by T-shirt stores.

Mobil Packaging, makers of Baggies, designed a better bag, a flat-bottomed model, and began trying to convert supermarkets to plastic bags back in 1975. Still no dice.

One factor continued to hold plastic back: price. But that changed at the 1985 New Materials and Profits in Grocery Sacks and Coextrusions Conference, where speaker Robert Bauman of Chem Systems announced that plastic bags were 11.5 percent cheaper than paper bags. You could hear the rush to the phones.

In case you missed it, *Plastics World* magazine revealed that plastic bags were in only 10 percent of the supermarkets in 1983. By the end of 1985 that had leaped to 75 percent. Today it is 88 percent and climbing rapidly.

And the reason for the switch is the same as the reason that held back plastic bags for years: price. The raw materials for plastic bags make up only about 50 percent of cost. For paper bags it is closer to 90 percent. The plastic-grocery-sack manufacturers can cut their price easily; the paper-sack people can't.

Today Judy picks plastic. That space between our refrigerator and the wall is already bulging with paper bags.

As the cashier drags each item across the X-shaped scanning table, a laser shoots a beam of light through a lens, which focuses it to a point. A wheel then sweeps that light-point across the bar code on the can or box. The beam is reflected back into the scanner's collection lens, where a photodetector reads it and sends an electric signal into a computer located in the store's office. The computer looks up the price in its master list and flashes it back to the cash register. It does all this in less than a second.

If the item is on sale, the computer has that sale price in its memory and it automatically rings up the sale price of the item. One computer can handle up to ninety cash registers. In addition to telling the cash register the price, the computer also logs the sale into its memory. If instructed, it can subtract the item from the store's inventory list and produce a report detailing the items that need to be restocked.

Scanning seems normal now. But when we first began shopping at this Winn-Dixie thirteen years ago, it was unheard of. Cashiers entered each item into the cash register.

Scanning was technologically feasible by 1972. But no supermarkets had scanners and no products had those little scanner codes. The Ad Hoc Committee on the Universal Product Code, a group composed of various members of the food-store chain—manufacturers, brokers, store owners—adopted the ten-digit UPC code that year. There was some debate over whether to print the code on the item label or to use an embossed or a magnetic strip. Manufacturers pushed for a printed code because it was less expensive, and the other parties went along.

The first supermarket scanner was put in operation on June 16, 1974, at Marsh Supermarket in Troy, Ohio. The first item scanned was a pack of Wrigley's Spearmint Chewing Gum. The system used an NCR cash register, computer, and scanner. It was no accident that Troy was selected for the scanner's debut. It was the home of Hobart, maker of UPC labels, and a half hour north of Dayton, home of National Cash Register.

That day 70 percent of the items in the store had a UPC code on them: Twenty-seven percent had the code printed on the label by the manufacturer; the other 43 percent had been tagged by store personnel. Today every item has the UPC code. In fact, the large supermarkets won't even accept a new product if it doesn't have that symbol.

But scanning's rapid rise hasn't been without controversy. Scanning was introduced as a cost-saving measure. Early estimates were that a store that did $140,000 a week in sales could save $110,000 a year. Of course, some of that savings was to come from no longer having to put a price tag on each item.

That hasn't happened, at least not in the numbers supermarket operators would like. Shoppers have resisted having to find the price on the shelf. In one early study customers were asked if they would accept having price tags taken off items if it meant substantial savings. One-fourth of the customers said the price should be left on regardless of the cost. Many stores were unwilling to alienate 25 percent of their customers, so they continued pricing individual items. The prospect of unmarked foodstuffs also got some legislators up in arms and a handful of states have mandatory price-marking laws.

Kentucky doesn't have such a law, so none of our groceries today has a price sticker.

Not that individual pricing was a feature in the old grocery stores of the thirties. It wasn't until the late forties that individual pricing became a standard. The first store to price each item was A&P. John Hartford, head of the A&P chain, stung by charges from the government's Antitrust Division that he was a "crooked merchant," decreed that every single item in his stores be marked with their retail price.

UPC symbols are a language unto themselves. But you can decipher them if you so desire. The two thin bars at the beginning and end are framing bars. Nothing more. The next bars on the left tell what kind of merchandise the item is. For instance, a wide space followed by a bar, a narrow space, and a thin bar is O, for groceries. The first five digits tell who the manufacturer is: 37000 is Procter & Gamble; 38000 is Kellogg's; 14100 is Pepperidge Farm. The next five digits are the product code. The last two digits are what's called a check digit, in case someone tries to alter the bar code with, say, a felt-tip pen. The check digits are a complex code—add up the numbers in the product code, divide by another number, add again; you get the idea— and if it doesn't match, the item won't scan. I don't know why someone would try to alter a UPC symbol anyway. Who's to say the new code might not be for a higher-priced item? The UPC bar code does not tell the price. That's in the computer.

The cashier manages to scan our 76 items in three minutes. A cashier should be able to average twenty items a minute. At many stores if a cashier falls much below that average, it's grounds for termination. Our cashier grips each item in the palm of her hand. It's what supermarkets call a power grip and most stores now instruct their cashiers to use this full-handed grip instead of a pinch grip with the fingers only. Studies have shown that cashiers who use a pinch grip are more likely to come down with repetitive stress syndrome.

As anyone who's ever emptied their grocery sacks to discover mashed potato chips and squashed bread knows, there's a right way and a wrong way to bag groceries. Our bag boy knows the right way. He's putting boxes and square items on the outer edges to make a liner form so the bag will stay square. Cans and other heavy items go in the middle. Lighter items like bread and chips go on top or in the space left between the heavier things. Baggers at many stores are told not to bag items with handles—

milk, orange juice—unless the customer asks them to. At Winn-Dixie they ask if we want the milk in a bag. We do.

Judy pays with a Visa credit card. It's not unusual for a supermarket to take credit cards, but it's still not the rule. Only 19

FORKLORE #36

IT WAS FORTY YEARS AGO TODAY

Food prices have climbed in the past forty years but not as much as you might think. Here are some food prices taken from my hometown newspaper on May 3, 1951.
   Cut-Rate Supermarket—"Open until 8 P.M. Fridays"—advertised:

1 lb. tin Van Camp's Chili and Beans—$0.33
1 lb. tin Van Camp's Beef Stew—$0.53
No. 2 1/2 tin Stokely's Fruit Cocktail—$0.33
No. 2 tin Del Monte Crushed Pineapple—$0.24
3 lb. tin Snowdrift shortening—$0.83
6-bottle carton Coca-Cola—$0.18
Half gallon Pet Ice Cream—$0.99
5 lb. tin Finger Lake Honey—$0.89
12 oz. jar JFG Smooth Peanut Butter—$0.29
1 lb. box Ritz Crackers—$0.29
Dozen Sunkist lemons—$0.29

That same day Oakwood Supermarket's ad featured:

Sauer's Pure Vanilla Extract—$0.14
No. 2 1/2 tin Bush's Pork and Beans—$0.25
4 oz. tin Armour Vienna Sausage—$0.23

A few pages later Kroger's ad spotlighted:

16 oz. can Kroger Pork and Beans—$0.10
Package of 10 Borden Biscuits—$0.10

percent of the nation's supermarkets accept credit cards.

We've had this Visa card since 1978, when we moved here and opened a bank account. It's hard to remember life before credit cards, when you actually had to pay cash for things. Credit cards are older than supermarkets. Gas credit cards were in use in the twenties. Department store credit cards came into use in the thirties. The first general-purpose credit cards were the Diner's Club cards, first issued in May 1950.

The cashier drags the card through a magnetic reader, which decodes the information on the card's magnetic strip: bank name, credit card number, type of account (Visa or MasterCard). The reader automatically dials the Visa banking center in Pittsburgh, which immediately records the total on our credit card account. The banking center computer, in turn, instructs the cash register to accept the charge and print out a credit card receipt.

Our Visa card account number has thirteen digits. Why so many numbers? It's not to deter thieves from memorizing the account number. It's to prevent clumsy clerks from typing in the wrong account number and charging our groceries to some poor schmo in Duluth. With thirteen digits there are 10 trillion possible card numbers. Should the clerk hit a wrong key, the chances of that mistyped number belonging to another card are one in 150,000.

Judy sighs at the sight of our bill: Her $2 worth of coupons brought it down to $174.88. It is a sigh that is part disappointment and part relief. "It's been more," she says. But when we were first married it was a lot less. Then, in 1977, you could figure about $5 a bag at the checkout. Now it's about $20 a bag.

When my mother charged her groceries at Joyner's store, my father would go down at the end of the month and pay the bill. It was something like twenty dollars a month. Say good-bye to the fifties.

As we begin our journey home, our coupons begin a journey of their own. They are bundled and shipped to a coupon center in Mexico, where laborers will tally them and send a check to Winn-

Dixie. It usually takes six to eight weeks from the time the coupons are mailed to the time the store is reimbursed.

The Statens may not be the all-American family, but we are very nearly the average American family: mom, dad, two children, two cars, and a weekly food bill that seldom drops below $100 (the average family of four spends $25 per person on food, but that includes eating out; for five or more that drops to $19 per person).

The Winn-Dixie computer has our name from the Visa card and our shopping habits from the scanner tally. It could pair these two together to track our shopping habits, our needs. And it could do this on every customer who pays with a credit card. If it wanted information on more of its customers, it could issue check-cashing cards or discount club cards and have a vast reservoir of customer information. A store could get instant results of an ad or a promotion or an in-store announcement. It could see how a price change affects sales. Winn-Dixie hasn't done that. Few supermarkets have.

It will happen though, as surely as shopping carts and self-service and scanners. Grocery stores will continue to take advantage of technology, and combining all of this scanner data in a computer is just another step forward. When they do, the supermarket will have come full circle from the days of the corner grocery. Because, in their own way, grocery stores will once again know their customers—if not personally, certainly one-by-one.

Index

Index

Index